Microbial Processing of Milk and Dairy Products

Microbial Processing of Milk and Dairy Products

Dr. Nand Lal Choudhary
Editor

KOROS PRESS LIMITED
London, UK

Microbial Processing of Milk and Dairy Products

© 2012

Printed in 2017 for Sale in the Indian Subcontinent

Published by
Koros Press Limited
3 The Pines, Rubery B45 9FF, Rednal,
Birmingham, United Kingdom

Tel.: +44-7826-930152
Email: info@korospress.com
www.korospress.com

ISBN: 978-1-78163-176-8

Editor: Dr. Nand Lal Choudhary

Printed in UK

British Library Cataloguing in Publication Data
A CIP record for this book is available from the British Library

10 9 8 7 6 5 4 3 2 1

No part of this publication may be reproduced, stored in a retrieval system or transmitted in any form or by any means, electronic, mechanical, photocopying, recording, scanning or otherwise without prior written permission of the publisher.

Reasonable efforts have been made to publish reliable data and information, but the authors, editors, and the publisher cannot assume responsibility for the legality of all materials or the consequences of their use. The authors, editors, and the publisher have attempted to trace the copyright holders of all materials in this publication and express regret to copyright holders if permission to publish has not been obtained. If any copyright material has not been acknowledged, let us know so we may rectify in any future reprint.

Exclusively distributed by CBS Publishers & Distributors Pvt. Ltd.
Sales & Distribution Rights only for India, Pakistan, Bangladesh, Sri Lanka, Nepal and Bhutan.This book is not to be sold outside these territories.

Contents

	Preface	*vii*
1.	**Microbiology of Dairy Products**	1

History • Dairy • Industrial Processing • Microbial Growth • Detection and Enumeration of Microorganisms • Microorganisms in Milk • Spoilage Microorganisms in Milk • Starter Cultures

2. **Controlling Microbial Quality of Dairy Products**   30

Recommended Minimum Internal Quality Control in Food Microbiology Testing Laboratories • Accreditation Scheme Requirements • An Overview of the Category Framework for Assessing Raw Milk Products • Application of Category Framework Approach • Bacteriological Tests for Milk • Milk Testing and Quality Control • Techniques Used in Milk Testing and Quality Control • Common Testing of Milk • Quality Control of Pasteurised Milk

3. **Physico-chemical Testing of Milk and Dairy Products**   50

Materials and Methods • Results and Discussion • Microbiological and Physicochemical Properties of Raw Milk Used for Processing Pasteurized Milk in Blue Nile Dairy Company (Sudan) • Results and Discussion • Hazard Analysis Critical Control Point—HACCP • History • The HACCP Seven Principles • HACCP Application • Emulsion • Shelf-life Predicting Methods for Milk • Determination of the End of Shelf Life for Milk • Materials and Methods • Results and Discussion • Indigenous (Indian) Dairy Products

4. **Chemistry of Dairy Products**   80

Physical Status of Milk • Milk Proteins • Milk Protein • Milk Protein Chemistry • Milk Protein Physical Properties • Deterioration of Milk Protein • Influence of Heat Treatment on Milk Proteins • Milk Processing • Fresh Milk Technology • Standardisation of Milk and Cream • Butter-making with Fresh Milk or Cream • Butter-making Theory • Churning Cream • Overrun and Produce in Butter-making • Butter Quality

- Ghee, Butter Oil and Dry Butterfat • Cheese-making
- Concentrated Fermented Milks • Sour-milk Technology
- Cottage Cheese

5. **Fluid Milk Processing** 127

   Beverage Milks • Concentrated and Dried Dairy Products • Dried Dairy Products • Cheese • Inoculation and Milk Ripening • Yogurt • Yogurt Products • Other Fermented Milk Beverages • Whipped Cream Structure • Ice Cream • Ice Cream: History and Folklore • Ice Cream Formulations • Formulation Considerations • Suggested Mixes • Ice Cream Ingredients • Lactose Crystallization • Mix Calculations for Ice Cream and Frozen Dairy Desserts • Ice Cream Manufacture • Ice Cream Novelty/Impulse Products • Ice Cream Flavours • Concentrated Extract • Fruit Ice Cream • Nuts in Ice Cream • Colour in Ice Cream • Homemade Ice Cream • Ingredients Used • Preparation of the Ice Cream Mix • Regular Vanilla Ice Cream • Milk and Cheese • Composition Of Milk • Chemical And Physical Properties • Homogenization of Milk • Reaction of Milk • Coagulation of Milk • Boiling and Heating of Milk • Sugar Reactions with Proteins of Milk • Cheese

6. **Heat Treatments and Pasteurization** 201

   The Purpose of Pasteurization • History of Pasteurization • Fluid Milk Production • Yogurt Production • Cheese Production • Ice Cream Production • Dairy Accounting • Milk Analysis • Estimation of Milk PH by Indicator • Determination of Milk Acidity • Alcohol Test • Cream • Determination of Milk Specific Gravity • Determination of moisture content of butter • Factors Affecting Milk Composition • Non-nutritional Factors Affecting Milk Composition • Nutritional Factors • Cow's Genetic Predisposition Affects Composition of Her Milk • Milk Composition • Fatty Acid Composition

7. **Chemicals: Lead, Mercury, Cadmium and Other Metals** 237

   Health Effects of Metals • Breast Milk • Benchmarks and Exposure Limits • Mercury • Minerals: Beneficial and Toxic • The Health Benefits of Raw Milk • A Word About Diet In General • Milk and Milk Products • Woman Milking a Goat • Standard Testing Procedures • Processing • Separation of Milk Components • Churning Cream • Working Butter with Butter Pats • Cultured/Fermented Dairy Products • Selling Curd from a Roadside Stall

   Bibliography 265

   Index 269

Preface

Whilst there are undoubtedly some who fear all microbes due to the association of some microbes with various human illnesses, many microbes are also responsible for numerous beneficial processes such as industrial fermentation, antibiotic production and as vehicles for cloning in more complex organisms such as plants. Scientists have also exploited their knowledge of microbes to produce biotechnological important enzymes such as Taq polymerase, reporter genes for use in other genetic systems and novel molecular biology techniques such as the yeast two-hybrid system. Bacteria can be used for the industrial production of amino acids. *Corynebacterium glutamicum* is one of the most important bacterial species with an annual production of more than two million tons of amino acids, mainly L-glutamate and L-lysine. A variety of biopolymers, such as polysaccharides, polyesters, and polyamides, are produced by microorganisms. Microorganisms are used for the biotechnological production of biopolymers with tailored properties suitable for high-value medical application such as tissue engineering and drug delivery. Microorganisms are used for the biosynthesis of xanthan, alginate, cellulose, cyanophycin, poly (gamma-glutamic acid), levan, hyaluronic acid, organic acids, oligosaccharides and polysaccharide, and polyhydroxyalkanoates.

Microorganisms are beneficial for microbial biodegradation or bioremediation of domestic, agricultural and industrial wastes and subsurface pollution in soils, sediments and marine environments. The ability of each microorganism to degrade toxic waste depends on the nature of each contaminant. Since sites typically have multiple pollutant types, the most effective approach to microbial biodegradation is to use a mixture of bacterial species and strains, each specific to the biodegradation of one or more types of contaminants.

There are also various claims concerning the contributions to human and animal health by consuming probiotics and/or prebiotics. Recent research has suggested that microorganisms could be useful in the treatment of cancer. Various strains of non-pathogenic clostridia can infiltrate and replicate within solid tumors. Clostridial vectors can be safely administered and their potential to deliver therapeutic proteins has been demonstrated in a variety of preclinical models.

A dairy is a building used for the harvesting of animal milk—mostly from cows or goats, but also from buffalo, sheep, horses or camels—for human consumption. A dairy is typically located on a dedicated *dairy farm* or section of a multi-purpose farm that is concerned with the harvesting of milk. Terminology differs between countries. For example, in the United States, a farm building where milk is harvesting is often called a *milking parlor*. In New Zealand such a building is historically know as the *milking shed* - although in recent years there has been a progressive change to call such a building a *farm dairy*.

The book is a unique resource for the students and teachers of food hygiene food safety and quality control in a wide range.

—*Dr. Nand Lal Choudhary*

1
Microbiology of Dairy Products

Microbiology is the study of *microorganisms*, which are microscopic, unicellular, and cell-cluster organisms. This includes eukaryotes such as fungi and protists, and prokaryotes. Viruses and prions, though not strictly classed as living organisms, are also studied. Microbiology typically includes the study of the immune system, or Immunology. Generally, immune systems interact with pathogenic microbes; these two disciplines often intersect which is why many colleges offer a paired degree such as "Microbiology and Immunology".

Microbiology is a broad term which includes virology, mycology, parasitology, bacteriology and other branches. A microbiologist is a specialist in microbiology and these other topics.

Microbiology is researched actively, and the field is advancing continually. It is estimated only about one percent of all of the microbe species on Earth have been studied. Although microbes were directly observed over three hundred years ago, the field of microbiology can be said to be in its infancy relative to older biological disciplines such as zoology and botany.

History

Ancient

The existence of microorganisms was hypothesized for many centuries before their actual discovery. The existence of unseen microbiological life was postulated by Jainism which is based on Mahavira's teachings as early as 6th century BCE.. Paul Dundas notes that Mahavira asserted existence of unseen microbiological creatures living in earth, water, air and fire. Jain scriptures also describe nigodas which are sub-microscopic creatures living in large clusters and having a very short life and are said to pervade each and every part of the universe, even in tissues of plants and flesh of animals. The Roman Marcus Terentius Varro made references to microbes when he warned against locating a homestead in the vicinity of swamps "because there are bred certain minute creatures which cannot be seen by the eyes, which float in the air and enter the body through the mouth and nose and there cause serious diseases."

In 1546 Girolamo Fracastoro proposed that epidemic diseases were caused by transferable seedlike entities that could transmit infection by direct or indirect contact, or even without contact over long distances.

However, early claims about the existence of microorganisms were speculative, and not based on any data or observation. Actual observation and discovery of microbes had to await the invention of the microscope in the 17th century.

Modern

In 1676, Antonie van Leeuwenhoek observed bacteria and other microorganisms, using a single-lens microscope of his own design. While Van Leeuwenhoek is often cited as the first to observe microbes, Robert Hooke made the first recorded microscopic observation, of the fruiting bodies of molds, in 1665.. The first observation of microbes using a microscope is generally credited to the Dutch draper and haberdasher, Antonie van Leeuwenhoek, who lived for most of his life in Delft, Holland. It has, however, been suggested that a Jesuit priest called Athanasius Kircher was the first to observe microorganisms.

He was among the first to design magic lanterns for projection purposes, so he must have been well acquainted with the properties of lenses. One of his book contains a chapter in Latin, which reads in translation – 'Concerning the wonderful structure of things in nature, investigated by Microscope. Here, he wrote 'who would believe that vinegar and milk abound with an innumerable multitude of worms.' He also noted that putrid material is full of innumerable creeping animalculae. These observations antedate Robert Hooke's Micrographia by nearly 20 years and were published some 29 years before van Leeuwenhoek saw protozoa and 37 years before he described having seen bacteria.

The field of bacteriology (later a subdiscipline of microbiology) was founded in the 19th century by Ferdinand Cohn, a botanist whose studies on algae and photosynthetic bacteria led him to describe several bacteria including *Bacillus* and *Beggiatoa*. Cohn was also the first to formulate a scheme for the taxonomic classification of bacteria and discover spores. Louis Pasteur and Robert Koch were contemporaries of Cohn's and are often considered to be the father of Microbiology and medical microbiology, respectively.

Pasteur is most famous for his series of experiments designed to disprove the then widely held theory of spontaneous generation, thereby solidifying microbiology's identity as a biological science. Pasteur also designed methods for food preservation (pasteurization) and vaccines against several diseases such as anthrax, fowl cholera and rabies.

Koch is best known for his contributions to the germ theory of disease, proving that specific diseases were caused by specific pathogenic microorganisms. He developed a series of criteria that have become known as the Koch's postulates. Koch was one of the first scientists to focus on the isolation of bacteria in pure culture resulting in his description of several novel bacteria including *Mycobacterium tuberculosis*, the causative agent of tuberculosis.

While Pasteur and Koch are often considered the founders of microbiology, their work did not accurately reflect the true diversity of the microbial world because of their exclusive focus on microorganisms having direct medical relevance. It was not until the late 19th century and the work of Martinus Beijerinck and Sergei Winogradsky, the founders of general microbiology (an older term encompassing aspects of microbial physiology, diversity and ecology), that the true breadth of microbiology was revealed. Beijerinck made two major contributions to microbiology: the discovery of viruses and the development of enrichment culture techniques.

While his work on the Tobacco Mosaic Virus established the basic principles of virology, it was his development of enrichment culturing that had the most immediate impact on microbiology by allowing for the cultivation of a wide range of microbes with wildly different physiologies. Winogradsky was the first to develop the concept of chemolithotrophy and to thereby reveal the essential role played by microorganisms in geochemical processes. He was responsible for the first isolation and description of both nitrifying and nitrogen-fixing bacteria.

Fields

The field of microbiology can be generally divided into several subdisciplines:

- Microbial physiology: The study of how the microbial cell functions biochemically. Includes the study of microbial growth, microbial metabolism and microbial cell structure.
- Microbial genetics: The study of how genes are organized and regulated in microbes in relation to their cellular functions. Closely related to the field of molecular biology.
- Cellular microbiology: A discipline bridging microbiology and cell biology.
- Medical microbiology: The study of the pathogenic microbes and the role of microbes in human illness. Includes the study of microbial pathogenesis and epidemiology and is related to the study of disease pathology and immunology.
- Veterinary microbiology: The study of the role in microbes in veterinary medicine or animal taxonomy.
- Environmental microbiology: The study of the function and diversity of microbes in their natural environments. Includes the study of microbial ecology, microbially-mediated nutrient cycling, geomicrobiology, microbial diversity and bioremediation. Characterization of key bacterial habitats such as the rhizosphere and phyllosphere, soil and ground-water ecosystems, open oceans or extreme environments (extremophiles).
- Evolutionary microbiology: The study of the evolution of microbes. Includes the study of bacterial systematics and taxonomy.
- Industrial microbiology: The exploitation of microbes for use in industrial processes. Examples include industrial fermentation and wastewater treatment. Closely linked to the biotechnology industry. This field also includes brewing, an important application of microbiology.
- Aeromicrobiology: The study of airborne microorganisms.
- Food microbiology: The study of microorganisms causing food spoilage and foodborne illness. Using microorganisms to produce foods, for example by fermentation.
- Pharmaceutical microbiology: the study of microorganisms causing pharmaceutical contamination and spoil
- Agricultural microbiology: The study of agriculturally important microorganisms.

(Jobs with the Centre For Disease Control and Prevention requires a degree in microbiology for most positions);
- Soil Microbiology: The study of those microorganisms that are found in soil.
- Water microbiology: The study of those microorganisms that are found in water.
- Generation microbiology: The study of those microorganisms that have the same characters as their parents.
- Nano microbiology: The study of those microorganisms at nano level.

Benefits

Whilst there are undoubtedly some who fear all microbes due to the association of some microbes with various human illnesses, many microbes are also responsible for numerous beneficial processes such as industrial fermentation (e.g. the production of alcohol, vinegar and dairy products), antibiotic production and as vehicles for cloning in more complex organisms such as plants. Scientists have also exploited their knowledge of microbes to produce biotechnological important enzymes such as Taq polymerase, reporter genes for use in other genetic systems and novel molecular biology techniques such as the yeast two-hybrid system.

Bacteria can be used for the industrial production of amino acids. *Corynebacterium glutamicum* is one of the most important bacterial species with an annual production of more than two million tons of amino acids, mainly L-glutamate and L-lysine. A variety of biopolymers, such as polysaccharides, polyesters, and polyamides, are produced by microorganisms. Microorganisms are used for the biotechnological production of biopolymers with tailored properties suitable for high-value medical application such as tissue engineering and drug delivery. Microorganisms are used for the biosynthesis of xanthan, alginate, cellulose, cyanophycin, poly (gamma-glutamic acid), levan, hyaluronic acid, organic acids, oligosaccharides and polysaccharide, and polyhydroxyalkanoates.

Microorganisms are beneficial for microbial biodegradation or bioremediation of domestic, agricultural and industrial wastes and subsurface pollution in soils, sediments and marine environments. The ability of each microorganism to degrade toxic waste depends on the nature of each contaminant. Since sites typically have multiple pollutant types, the most effective approach to microbial biodegradation is to use a mixture of bacterial species and strains, each specific to the biodegradation of one or more types of contaminants.

There are also various claims concerning the contributions to human and animal health by consuming probiotics (bacteria potentially beneficial to the digestive system) and/or prebiotics (substances consumed to promote the growth of probiotic microorganisms).

Recent research has suggested that microorganisms could be useful in the treatment of cancer. Various strains of non-pathogenic clostridia can infiltrate and replicate within solid tumors. Clostridial vectors can be safely administered and their potential to deliver therapeutic proteins has been demonstrated in a variety of preclinical models.

Dairy

A dairy is a building used for the harvesting of animal milk—mostly from cows or goats, but also from buffalo, sheep, horses or camels —for human consumption. A dairy is typically located on a dedicated *dairy farm* or section of a multi-purpose farm that is concerned with the harvesting of milk.

Terminology differs between countries. For example, in the United States, a farm building where milk is harvesting is often called a *milking parlor*. In New Zealand such a building is historically know as the *milking shed* - although in recent years there has been a progressive change to call such a building a *farm dairy*.

In some countries, especially those with small numbers of animals being milked, as well as harvesting the milk from an animal, the dairy may also process the milk into butter, cheese and yoghurt, for example. This is a traditional method of producing specialist milk products, especially in Europe. In the United States a *dairy* can also be a place that processes, distributes and sells dairy products, or a room, building or establishment where milk is stored and processed into milk products, such as butter or cheese. In New Zealand English the singular use of the word *dairy* almost exclusively refers to the corner convenience store, or superette. This usage is historical as such stores were a common place for the public to buy milk products. As an attributive, the word *dairy* refers to milk-based products, derivatives and processes, and the animals and workers involved in their production: for example dairy cattle, dairy goat. A dairy farm produces milk and a dairy factory processes it into a variety of dairy products. These establishments constitute the dairy industry, a component of the food industry.

History

Milk producing animals have been domesticated for thousands of years. Initially, they were part of the subsistence farming that nomads engaged in. As the community moved about the country, their animals accompanied them. Protecting and feeding the animals were a big part of the symbiotic relationship between the animals and the herders.

In the more recent past, people in agricultural societies owned dairy animals that they milked for domestic and local (village) consumption, a typical example of a cottage industry. The animals might serve multiple purposes (for example, as a draught animal for pulling a plough as a youngster, and at the end of its useful life as meat). In this case the animals were normally milked by hand and the herd size was quite small, so that all of the animals could be milked in less than an hour—about 10 per milker. These tasks were performed by a *dairymaid* (*dairywoman*) or *dairyman*. The word *dairy* harkens back to Middle English *dayerie*, *deyerie*, from *deye* (female servant or dairymaid) and further back to Old English *dæge* (kneader of bread).

With industrialisation and urbanisation, the supply of milk became a commercial industry, with specialised breeds of cattle being developed for dairy, as distinct from beef or draught animals. Initially, more people were employed as milkers, but it soon turned to mechanisation with machines designed to do the milking.

Historically, the milking and the processing took place close together in space and time: on a dairy farm. People milked the animals by hand; on farms where only small numbers are kept, hand-milking may still be practiced. Hand-milking is accomplished by grasping the teats (often pronounced *tit* or *tits*) in the hand and expressing milk either by squeezing the fingers progressively, from the udder end to the tip, or by squeezing the teat between thumb and index finger, then moving the hand downward from udder towards the end of the teat. The action of the hand or fingers is designed to close off the milk duct at the udder (upper) end and, by the movement of the fingers, close the duct progressively to the tip to express the trapped milk. Each half or quarter of the udder is emptied one milk-duct capacity at a time.

The *stripping* action is repeated, using both hands for speed. Both methods result in the milk that was trapped in the milk duct being squirted out the end into a bucket that is supported between the knees (or rests on the ground) of the milker, who usually sits on a low stool. Traditionally the cow, or cows, would stand in the field or paddock while being milked. Young stock, heifers, would have to be trained to remain still to be milked. In many countries, the cows were tethered to a post and milked. The problem with this method is that it relies on quiet, tractable beasts, because the hind end of the cow is not restrained.

In 1937, it was found that bovine somatotropin (bST or bovine growth hormone) would increase the yield of milk. Monsanto Company developed a synthetic (recombinant) version of this hormone (rBST). In February 1994, rBST was approved by the Food and Drug Administration (FDA) for use in the U.S. It has become common in the U.S., but not elsewhere, to inject it into milch kine dairy cows to increase their production by up to 15%.

However, there are claims that this practice can have negative consequences for the animals themselves. A European Union scientific commission was asked to report on the incidence of mastitis and other disorders in dairy cows, and on other aspects of the welfare of dairy cows. The commission's statement, subsequently adopted by the European Union, stated that the use of rBST substantially increased health problems with cows, including foot problems, mastitis and injection site reactions, impinged on the welfare of the animals and caused reproductive disorders. The report concluded that on the basis of the health and welfare of the animals, rBST should not be used. Health Canada prohibited the sale of rBST in 1999; the recommendations of external committees were that, despite not finding a significant health risk to humans, the drug presented a threat to animal health and, for this reason, could not be sold in Canada.

Structure of the Industry

While most countries produce their own milk products, the structure of the dairy industry varies in different parts of the world. In major milk-producing countries most milk is distributed through wholesale markets. In Ireland and Australia, for example, farmers' co-operatives own many of the large-scale processors, while in the United States many farmers and processors do business through individual contracts. In the United States, the country's 196 farmers' cooperatives sold 86% of milk in the U.S. in 2002, with five cooperatives accounting for half that. This was down from 2,300 cooperatives in the 1940s. In developing countries, the past practice of farmers marketing milk in their own neighbourhoods are changing rapidly. Notable developments include considerable foreign investment in the dairy industry and a growing role for dairy cooperatives. Output of milk is growing rapidly in such countries and presents a major source of income growth for many farmers.

As in many other branches of the food industry, dairy processing in the major dairy producing countries has become increasingly concentrated, with fewer but larger and more efficient plants operated by fewer workers. This is notably the case in the United States, Europe, Australia and New Zealand. In 2009, charges of anti-trust violations have been made against major dairy industry players in the United States.

Government intervention in milk markets was common in the 20th century. A limited anti-trust exemption was created for U.S. dairy cooperatives by the Capper-Volstead Act of 1922. In the 1930s, some U.S. states adopted price controls, and Federal Milk Marketing Orders started under the Agricultural Marketing Agreement Act of 1937 and continue in the 2000s. The Federal Milk Price Support Program began in 1949. The Northeast Dairy Compact regulated wholesale milk prices in New England from 1997 to 2001.

Plants producing liquid milk and products with short shelf life, such as yogurts, creams and soft cheeses, tend to be located on the outskirts of urban centres close to consumer markets. Plants manufacturing items with longer shelf life, such as butter, milk powders, cheese and whey powders, tend to be situated in rural areas closer to the milk supply. Most large processing plants tend to specialise in a limited range of products. Exceptionally, however, large plants producing a wide range of products are still common in Eastern Europe, a holdover from the former centralized, supply-driven concept of the market.

As processing plants grow fewer and larger, they tend to acquire bigger, more automated and more efficient equipment. While this technological tendency keeps manufacturing costs lower, the need for long-distance transportation often increases the environmental impact.

Milk production is irregular, depending on cow biology. Producers must adjust the mix of milk which is sold in liquid form vs. processed foods (such as butter and cheese) depending on changing supply and demand.

Operation of the Dairy Farm

When it became necessary to milk larger numbers of cows, the cows would be brought to a shed or barn that was set up with bails (stalls) where the cows could be confined while they were milked. One person could milk more cows this way, as many as 20 for a skilled worker. But having cows standing about in the yard and shed waiting to be milked is not good for the cow, as she needs as much time in the paddock grazing as is possible. It is usual to restrict the twice-daily milking to a maximum of an hour and a half each time. It makes no difference whether one milks 10 or 1000 cows, the milking time should not exceed a total of about three hours each day for any cow.

As herd sizes increased there was more need to have efficient milking machines, sheds, milk-storage facilities (vats), bulk-milk transport and shed cleaning capabilities and the means of getting cows from paddock to shed and back.

Farmers found that cows would abandon their grazing area and walk towards the milking area when the time came for milking. This is not surprising as, in the flush of the milking season, cows presumably get very uncomfortable with udders engorged with milk, and the place of relief for them is the milking shed.

As herd numbers increased so did the problems of animal health. In New Zealand two approaches to this problem have been used. The first was improved veterinary medicines (and the government regulation of the medicines) that the farmer could use. The other was the creation of *veterinary clubs* where groups of farmers would employ a veterinarian (vet) full-time and share those services throughout the year. It was in the vet's interest to keep the animals healthy and reduce the number of calls from farmers, rather than to ensure that the farmer needed to call for service and pay regularly. Most dairy farmers milk their cows with absolute regularity at a minimum of twice a day, with some high-producing herds milking up to four times a day to lessen the weight of large volumes of milk in the udder of the cow. This daily milking routine goes on for about 300 to 320 days per year that the cow stays in milk. Some small herds are milked once a day for about the last 20 days of the production cycle but this is not usual for large herds.

If a cow is left unmilked just once she is likely to reduce milk-production almost immediately and the rest of the season may see her *dried off* (giving no milk) and still consuming feed for no production. However, once-a-day milking is now being practised more widely in New Zealand for profit and lifestyle reasons. This is effective because the fall in milk yield is at least partially offset by labour and cost savings from milking once per day. This compares to some intensive farm systems in the United States that milk three or more times per day due to higher milk yields per cow and lower marginal labour costs.

Farmers who are contracted to supply liquid milk for human consumption (as opposed to milk for processing into butter, cheese, and so on—see milk) often have to manage their herd so that the contracted number of cows are in milk the year round, or the required minimum milk output is maintained. This is done by mating cows outside their natural mating time so that the period when each cow in the herd is giving maximum production is in rotation throughout the year.

Northern hemisphere farmers who keep cows in barns almost all the year usually manage their herds to give continuous production of milk so that they get paid all year round. In the southern hemisphere the cooperative dairying systems allow for two months on no productivity because their systems are designed to take advantage of maximum grass and milk production in the spring and because the milk processing plants pay bonuses in the dry (winter) season to carry the farmers through the mid-winter break from milking. It also means that cows have a rest from milk production when they are most heavily pregnant. Some year-round milk farms are penalised financially for over-production at any time in the year by being unable to sell their overproduction at current prices.

Artificial insemination (AI) is common in all high-production herds.

Industrial Processing

Dairy plants process the raw milk they receive from farmers so as to extend its marketable life. Two main types of processes are employed: heat treatment to ensure the safety of milk for human consumption and to lengthen its shelf-life, and dehydrating dairy products such as butter, hard cheese and milk powders so that they can be stored.

Cream and Butter

Today, milk is separated by large machines in bulk into cream and skim milk. The cream is processed to produce various consumer products, depending on its thickness, its suitability for culinary uses and consumer demand, which differs from place to place and country to country. Some cream is dried and powdered, some is condensed (by evaporation) mixed with varying amounts of sugar and canned. Most cream from New Zealand and Australian factories is made into butter. This is done by churning the cream until the fat globules coagulate and form a monolithic mass. This butter mass is washed and, sometimes, salted to improve keeping qualities. The residual buttermilk goes on to further processing. The butter is packaged (25 to 50 kg boxes) and chilled for storage and sale. At a later stage these packages are broken down into home-consumption sized packs.

Skimmed Milk

The product left after the cream is removed is called skim, or skimmed, milk. To make a consumable liquid a portion of cream is returned to the skim milk to make *low fat milk* (semi-skimmed) for human consumption. By varying the amount of cream returned, producers can make a variety of low-fat milks to suit their local market. Other products, such as calcium, vitamin D, and flavouring, are also added to appeal to consumers.

Casein

Casein is the predominant phosphoprotein found in fresh milk. It has a very wide range of uses from being a filler for human foods, such as in ice cream, to the manufacture of products such as fabric, adhesives, and plastics.

Cheese

Cheese is another product made from milk. Whole milk is reacted to form curds that can be compressed, processed and stored to form cheese. In countries where milk is legally allowed to be processed without pasteurisation a wide range of cheeses can be made using the bacteria naturally in the milk. In most other countries, the range of cheeses is smaller and the use of artificial cheese curing is greater. Whey is also the by-product of this process.

Whey

In earlier times whey was considered to be a waste product and it was, mostly, fed to pigs as a convenient means of disposal. Beginning about 1950, and mostly since about 1980, lactose and many other products, mainly food additives, are made from both casein and cheese whey.

Yogurt

Yoghurt (or yogurt) making is a process similar to cheese making, only the process is arrested before the curd becomes very hard.

Milk Powders

Milk is also processed by various drying processes into powders. Whole milk, skim milk, buttermilk, and whey products are dried into a powder form and used for human and animal consumption. The main difference between production of powders for human or for animal consumption is in the protection of the process and the product from contamination. Some people drink milk reconstituted from powdered milk, because milk is about 88% water and it is much cheaper to transport the dried product.

Other Milk Products

Kumis is produced commercially in Central Asia. Although it is traditionally made from mare's milk, modern industrial variants may use cow's milk instead.

Transport of Milk

Historically, the milking and the processing took place in the same place: on a dairy farm. Later, cream was separated from the milk by machine, on the farm, and the cream was transported to a factory for butter making. The skim milk was fed to pigs. This allowed for the high cost of transport (taking the smallest volume high-value product), primitive trucks and the poor quality of roads. Only farms close to factories could afford to take whole milk, which was essential for cheese making in industrial quantities, to them. The development of refrigeration and better road transport, in the late 1950s, has meant that most farmers milk their cows and only temporarily store the milk in large refrigerated bulk tanks, from where it is later transported by truck to central processing facilities.

Milking Machines

Milking machines are used to harvest milk from cows when manual milking becomes inefficient or labour intensive. The milking unit is the portion of a milking machine for removing milk from an udder. It is made up of a claw, four teatcups, (Shells and rubber liners) long milk tube, long pulsation tube, and a pulsator. The claw is an assembly that connects the short pulse tubes and short milk tubes from the teatcups to the long pulse tube and long milk tube. (Cluster assembly) Claws are commonly made of stainless steel or plastic or both. Teatcups are composed of a rigid outer shell (stainless steel or plastic) that holds a soft inner liner or *inflation*. Transparent sections in the shell may allow viewing of liner collapse and milk flow. The annular space between the shell and liner is called the pulse chamber.

Milking machines work in a way that is different from hand milking or calf suckling. Continuous vacuum is applied inside the soft liner to massage milk from the teat by creating a pressure difference across the teat canal (or opening at the end of the teat). Vacuum also helps keep the machine attached to the cow. The vacuum applied to the teat causes congestion of teat tissues (accumulation of blood and other fluids). Atmospheric air is admitted into the pulsation chamber about once per second (the pulsation rate) to allow the liner to collapse around the end of teat and relieve congestion in the teat tissue. The ratio of the time that the liner is open (milking phase) and closed (rest phase) is called the pulsation ratio.

The four streams of milk from the teatcups are usually combined in the claw and transported to the milkline, or the collection bucket (usually sized to the output of one cow) in a single milk hose. Milk is then transported (manually in buckets) or with a combination of airflow and mechanical pump to a central storage vat or bulk tank. Milk is refrigerated on the farm in most countries either by passing through a heat-exchanger or in the bulk tank, or both.

In the photo above is a bucket milking system with the stainless steel bucket visible on the far side of the cow. The two rigid stainless steel teatcup shells applied to the front two quarters of the udder are visible. The top of the flexible liner is visible at the top of the shells as are the short milk tubes and short pulsation tubes extending from the bottom of the shells to the claw. The bottom of the claw is transparent to allow observation of milk flow. When milking is completed the vacuum to the milking unit is shut off and the teatcups are removed.

Milking machines keep the milk enclosed and safe from external contamination. The interior 'milk contact' surfaces of the machine are kept clean by a manual or automated washing procedures implemented after milking is completed. Milk contact surfaces must comply with regulations requiring food-grade materials (typically stainless steel and special plastics and rubber compounds) and are easily cleaned.

Most milking machines are powered by electricity but, in case of electrical failure, there can be an alternative means of motive power, often an internal combustion engine, for the vacuum and milk pumps. Milk cows cannot tolerate delays in scheduled milking without serious milk production reductions.

Milking Shed Layouts

Bail-style sheds— This type of milking facility was the first development, after open-paddock milking, for many farmers. The building was a long, narrow, *lean-to* shed that was open along one long side. The cows were held in a yard at the open side and when they were about to be milked they were positioned in one of the bails (stalls). Usually the cows were restrained in the bail with a breech chain and a rope to restrain the outer back leg. The cow could not move about excessively and the milker could expect not to be kicked or trampled while sitting on a (three-legged) stool and milking into a bucket. When each cow was finished she backed out into the yard again. The UK bail, developed largely by Rex Patterson, was a six standing mobile shed with steps that the cow mounted, so the herdsman didn't have to bend so low. The milking equipment was much as today, a vacuum from a pump, pulsators, a claw-piece with pipes leading to the four shells and liners that stimulate and suck the milk from the teat. The milk went into churns, via a cooler.

As herd sizes increased a door was set into the front of each bail so that when the milking was done for any cow the milker could, after undoing the leg-rope and with a remote link, open the door and allow her to exit to the pasture. The door was closed, the next cow walked into the bail and was secured. When milking machines were introduced bails were set in pairs so that a cow was being milked in one paired bail while the other could be prepared for milking. When one was finished the machine's cups are swapped to the other cow. This is the same as for *Swingover Milking Parlours* as described below except that the cups are loaded on the udder from the side. As herd numbers increased it was easier to double-up the cup-sets and milk both cows simultaneously than to increase the number of bails. About 50 cows an hour can be milked in a shed with 8 bales by one person. Using the same teat cups for successive cows has the danger of transmitting infection, mastitis, from one cow to another. Some farmers have devised their own ways to disinfect the clusters between cows.

Herringbone Milking Parlours— In herringbone milking sheds, or parlours, cows enter, in single file, and line up almost perpendicular to the central aisle of the milking parlour on both sides of a central pit in which the milker works (you can visualise a fishbone with the ribs representing the cows and the spine being the milker's working area; the cows face outward). After washing the udder and teats the cups of the milking machine are applied to the cows, from the rear of their hind legs, on both sides of the working area. Large herringbone sheds can milk up to 600 cows efficiently with two people.

Swingover Milking Parlours— Swingover parlours are the same as herringbone parlours except they have only one set of milking cups to be shared between the two rows of cows, as one side is being milked the cows on the other side are moved out and replaced with unmilked ones. The advantage of this system is that it is less costly to equip, however it operates at slightly better than half-speed and one would not normally try to milk more than about 100 cows with one person.

Rotary Milking sheds— Rotary milking sheds consist of a turntable with about 12 to 100 individual stalls for cows around the outer edge. A "good" rotary will be operated with 24–32 (~48–50+) stalls by one (two) milkers. The turntable is turned by an electric-motor drive at a rate that one turn is the time for a cow to be milked completely. As an empty stall passes the entrance a cow steps on, facing the centre, and rotates with the turntable.

The next cow moves into the next vacant stall and so on. The operator, or milker, cleans the teats, attaches the cups and does any other feeding or whatever husbanding operations that are necessary. Cows are milked as the platform rotates. The milker, or an automatic device, removes the milking machine cups and the cow backs out and leaves at an exit just before the entrance. The rotary system is capable of milking very large herds—over a thousand cows.

Automatic Milking sheds— Automatic milking or 'robotic milking' sheds can be seen in Australia, New Zealand and many European countries. Current automatic milking sheds use the voluntary milking (VM) method. These allow the cows to voluntarily present themselves for milking at any time of the day or night, although repeat visits may be limited by the farmer through computer software. A robot arm is used to clean teats and apply milking equipment, while automated gates direct cow traffic, eliminating the need for the farmer to be present during the process. The entire process is computer controlled.

Supplementary accessories in sheds— Farmers soon realised that a milking shed was a good place to feed cows supplementary foods that overcame local dietary deficiencies or added to the cows' wellbeing and production. Each bail might have a box into which such feed is delivered as the cow arrives so that she is eating while being milked. A computer can read the eartag of each beast to ration the correct individual supplement. A close alternative is to use 'out-of-parlour-feeders', stalls that respond to a transponder around the cow's neck that is programmed to provide each cow with a supplementary feed, the quantity dependent on her production, stage in lactation, and the benefits of the main ration.

The holding yard at the entrance of the shed is important as a means of keeping cows moving into the shed. Most yards have a powered gate that ensures that the cows are kept close to the shed.

Water is a vital commodity on a dairy farm: cows drink about 20 gallons (80 litres) a day, sheds need water to cool and clean them. Pumps and reservoirs are common at milking facilities. Water can be warmed by heat transfer with milk.

Temporary Milk Storage

Milk coming from the cow is transported to a nearby storage vessel by the airflow leaking around the cups on the cow or by a special "air inlet" (5-10 l/min free air) in the claw. From there it is pumped by a mechanical pump and cooled by a heat exchanger. The milk is then stored in a large vat, or bulk tank, which is usually refrigerated until collection for processing.

Processing Facilities

Topics:
- Pasteurization, homogenization
- Cream extraction
- Cheese making
- Butter making
- Caseinmaking
- Yogurt processing.

Waste Disposal

In countries where cows are grazed outside year-round, there is little waste disposal to deal with. The most concentrated waste is at the milking shed, where the animal waste is liquefied (during the water-washing process) and allowed to flow by gravity, or pumped, into composting ponds with anaerobic bacteria to consume the solids. The processed water and nutrients are then pumped back onto the pasture as irrigation and fertilizer. Surplus animals are slaughtered for processed meat and other rendered products.

In the associated milk processing factories, most of the waste is washing water that is treated, usually by composting, and returned to waterways. This is much different from half a century ago, when the main products were butter, cheese and casein, and the rest of the milk had to be disposed of as waste (sometimes as animal feed).

In areas where cows are housed all year round, the waste problem is difficult because of the amount of feed that is brought in and the amount of bedding material that also has to be removed and composted. The size of the problem can be understood by standing downwind of the barns where such dairying goes on.

In many cases, modern farms have very large quantities of milk to be transported to a factory for processing. If anything goes wrong with the milking, transport or processing facilities it can be a major disaster trying to dispose of enormous quantities of milk. If a road tanker overturns on a road, the rescue crew is looking at accommodating the spill of 5 to 10 thousand gallons of milk (20 to 45 thousand litres) without allowing any into the waterways.

A derailed rail tanker-train may involve 10 times that amount. Without refrigeration, milk is a fragile commodity, and it is very damaging to the environment in its raw state due to its high biochemical oxygen demand. A widespread electrical power blackout is another disaster for the dairy industry, because both milking and processing facilities are affected. For this, farms may often use mobile generators. Such a situation occurred during the power outage caused by the 2010 Canterbury Earthquake.

In dairy-intensive areas, various methods have been proposed for disposing of large quantities of milk. These directives include feeding milk to livestock, spray irrigation or designating a sacrifice area. Large application rates of milk onto land, or disposing in a hole, is problematic as the residue from the decomposing milk will block the soil pores and thereby reduce the water infiltration rate through the soil profile. As recovery of this effect can take time, any land based application needs to be well managed and considered.

Associated Diseases

- Leptospirosis is one of the most common debilitating diseases of milkers, made somewhat worse since the introduction of herringbone sheds, because of unavoidable direct contact with bovine urine
- Cowpox is one of the helpful diseases; it is barely harmful to humans and tends to inoculate them against other poxes such as small pox.
- Tuberculosis (TB) is able to be transmitted from cattle mainly via milk products that are unpasteurised. TB has been eradicated from many countries by testing for the disease and culling suspected animals.
- Brucellosis is a bacterial disease transmitted to humans by dairy products and direct animal contact. Brucellosis has been eradicated from certain countries by testing for the disease and culling suspected animals
- Listeria is a bacterial disease associated with unpasteurised milk, and can affect some cheeses made in traditional ways. Careful observance of the traditional cheese making methods achieves reasonable protection for the consumer.
- Johne's Disease (pronounced "yo-knees") is a contagious, chronic and sometimes fatal infection in ruminants caused by a bacterium named Mycobacterium avium subspecies paratuberculosis (M. paratuberculosis). The bacteria are present in retail milk, and are believed by some researchers to be the primary cause of Crohn's disease in humans. This disease is not known to infect animals in Australia and New Zealand.

Microorganisms

Microorganisms are living organisms that are individually too small to see with the naked eye. The unit of measurement used for microorganisms is the micrometer (μ m); 1 μ m = 0.001 millimeter; 1 nanometer (nm) = 0.001 μ m. Microorganisms are found everywhere (ubiquitous) and are essential to many of our planets life processes. With regards to the food industry, they can cause spoilage, prevent spoilage through fermentation, or can be the cause of human illness.

There are several classes of microorganisms, of which bacteria and fungi (yeasts and moulds) will be discussed in some detail. Another type of microorganism, the bacterial viruses or bacteriophage, will be examined in a later section.

Bacteria

Bacteria are relatively simple single-celled organisms. One method of classification is by shape or morphology:

- Cocci:
  — spherical shape
  — 0.4 - 1.5 µ m

Examples: staphylococci - form grape-like clusters; streptococci - form bead-like chains;

- Rods:
  — 0.25 - 1.0 µ m width by 0.5 - 6.0 µ m long.

Examples: bacilli - straight rod; spirilla - spiral rod.

There exists a bacterial system of taxonomy, or classification system, that is internationally recognized with family, genera and species divisions based on genetics.

Some bacteria have the ability to form resting cells known as endospores. The spore forms in times of environmental stress, such as lack of nutrients and moisture needed for growth, and thus is a survival strategy. Spores have no metabolism and can withstand adverse conditions such as heat, disinfectants, and ultraviolet light. When the environment becomes favourable, the spore germinates and giving rise to a single vegetative bacterial cell. Some examples of spore-formers important to the food industry are members of *Bacillus* and *Clostridium* generas.

Bacteria reproduce asexually by fission or simple division of the cell and its contents. The doubling time, or generation time, can be as short as 20-20 min. Since each cell grows and divides at the same rate as the parent cell, this could under favourable conditions translate to an increase from one to 10 million cells in 11 hours! However, bacterial growth in reality is limited by lack of nutrients, accumulation of toxins and metabolic wastes, unfavourable temperatures and dessication. The maximum number of bacteria is approximately $1 \times 10e9$ CFU/g or ml.

Note: Bacterial populations are expressed as colony forming units (CFU) per gram or millilitre.

Bacterial growth generally proceeds through a series of phases:
- Lag phase: time for microorganisms to become accustomed to their new environment. There is little or no growth during this phase.

- Log phase: bacteria logarithmic, or exponential, growth begins; the rate of multiplication is the most rapid and constant.
- Stationary phase: the rate of multiplication slows down due to lack of nutrients and build-up of toxins. At the same time, bacteria are constantly dying so the numbers actually remain constant.
- Death phase: cell numbers decrease as growth stops and existing cells die off.

The shape of the curve varies with temperature, nutrient supply, and other growth factors. This exponential death curve is also used in modeling the heating destruction of microorganisms.

Yeasts

Yeasts are members of a higher group of microorganisms called fungi. They are single-cell organisms of spherical, elliptical or cylindrical shape. Their size varies greatly but are generally larger than bacterial cells. Yeasts may be divided into two groups according to their method of reproduction:

1. budding: called Fungi Imperfecti or false yeasts
2. budding and spore formation: called Ascomycetes or true yeasts.

Unlike bacterial spores, yeast form spores as a method of reproduction.

Moulds

Moulds are filamentous, multi-celled fungi with an average size larger than both bacteria and yeasts (10 X 40 µ m). Each filament is referred to as a hypha. The mass of hyphae that can quickly spread over a food substrate is called the mycelium. Moulds may reproduce either asexually or sexually, sometimes both within the same species.

Asexual Reproduction:
- fragmentation - hyphae separate into individual cells called arthropsores
- spore production - formed in the tip of a fruiting hyphae, called conidia, or in swollen structures called sporangium.

Sexual Reproduction: sexual spores are produced by nuclear fission in times of unfavourable conditions to ensure survival.

Microbial Growth

There are a number of factors that affect the survival and growth of microorganisms in food. The parameters that are inherent to the food, or intrinsic factors, include the following:

- nutrient content
- moisture content
- pH
- available oxygen
- biological structures
- antimicrobial constituents.

Nutrient Requirements: While the nutrient requirements are quite organism specific, the microorganisms of importance in foods require the following:
- water
- energy source
- carbon/nitrogen source
- vitamins
- minerals.

Milk and dairy products are generally very rich in nutrients which provides an ideal growth environment for many microorganisms.

Moisture Content: All microorganisms require water but the amount necessary for growth varies between species. The amount of water that is available in food is expressed in terms of water activity (aw), where the aw of pure water is 1.0. Each microorganism has a maximum, optimum, and minimum aw for growth and survival. Generally bacteria dominate in foods with high aw (minimum approximately 0.90 aw) while yeasts and moulds, which require less moisture, dominate in low aw foods (minimum 0.70 aw). The water activity of fluid milk is approximately 0.98 aw.

pH: Most microorganisms have approximately a neutral pH optimum (pH 6-7.5). Yeasts are able to grow in a more acid environment compared to bacteria. Moulds can grow over a wide pH range but prefer only slightly acid conditions. Milk has a pH of 6.6 which is ideal for the growth of many microoorganisms.

Available Oxygen: Microorganisms can be classified according to their oxygen requirements necessary for growth and survival:
- Obligate Aerobes: oxygen required
- Facultative: grow in the presence or absence of oxygen
- Microaerophilic: grow best at very low levels of oxygen
- Aerotolerant Anaerobes: oxygen not required for growth but not harmful if present

- Obligate Anaerobes: grow only in complete absence of oxygen; if present it can be lethal.

Biological Structures: Physical barriers such as skin, rinds, feathers, etc. have provided protection to plants and animals against the invasion of microorganisms. Milk, however, is a fluid product with no barriers to the spreading of microorganisms throughout the product.

Antimicrobial Constituents: As part of the natural protection against microorganisms, many foods have antimicrobial factors. Milk has several nonimmunological proteins which inhibit the growth and metabolism of many microorganisms including the following most common:

1. lactoperoxidase
2. lactoferrin
3. lysozyme
4. xanthine.

Where the intrinsic factors are related to the food properties, the extrinsic factors are related to the storage environment. These would include temperature, relative humidity, and gases that surround the food.

Temperature: As a group, microorganisms are capable of growth over an extremely wide temperature range. However, in any particular environment, the types and numbers of microorganisms will depend greatly on the temperature. According to temperature, microorganisms can be placed into one of three broad groups:

- Psychrotrophs: optimum growth temperatures 20 to 30° capable of growth at temperatures less than 7° C. Psychrotrophic organisms are specifically important in the spoilage of refrigerated dairy products.
- Mesophiles: optimum growth temperatures 30 to 40° C; do not grow at refrigeration temperatures
- Thermophiles: optimum growth between 55 and 65° C

It is important to note that for each group, the growth rate increases as the temperature increases only up to an optimum, afterwhich it rapidly declines.

Detection and Enumeration of Microorganisms

There are several methods for detection and enumeration of microorganisms in food. The method that is used depends on the purpose of the testing.

Direct Enumeration

Using direct microscopic counts (DMC), Coulter counter etc. allows a rapid estimation of all viable and nonviable cells. Identification through staining and observation of morphology also possible with DMC.

Viable Enumeration

The use of standard plate counts, most probable number (MPN), membrane filtration, plate loop methos, spiral plating etc., allows the estimation of only viable cells. As with direct enumeration, these methods can be used in the food industry to enumerate fermentation, spoilage, pathogenic, and indicator organisms.

Metabolic Activity Measurement

An estimation of metabolic activity of the total cell population is possible using dye reduction tests such as resazurin or methylene blue dye reduction, acid production, electrical impedence etc. The level of bacterial activity can be used to assess the keeping quality and freshness of milk. Toxin levels can also be measured, indicating the presence of toxin producing pathogens.

Cellular Constituents Measurement

Using the luciferase test to measure ATP is one example of the rapid and sensitive tests available that will indicate the presence of even one pathogenic bacterial cell.

Isolation of microorganisms is an important preliminary step in the identification of most food spoilage and pathogenic organisms. This can be done using a simple streak plate method.

Microorganisms in Milk

Milk is sterile at secretion in the udder but is contaminated by bacteria even before it leaves the udder. Except in the case of mastisis, the bacteria at this point are harmless and few in number. Further infection of the milk by microorganisms can take place during milking, handling, storage, and other pre-processing activities.

Lactic acid bacteria: this group of bacteria are able to ferment lactose to lactic acid. They are normally present in the milk and are also used as starter cultures in the production of cultured dairy products such as yogurt. Note: many lactic acid bacteria have recently been reclassified; the older names will appear in brackets as you will

still find the older names used for convenience sake in a lot of literature. Some examples in milk are:
- lactococci
  - *L. delbrueckii* subsp. *lactis* (*Streptococcus lactis*)
  - *Lactococcus lactis* subsp. *cremoris* (*Streptococcus cremoris*)
- lactobacilli
  - *Lactobacillus casei*
  - *L.delbrueckii* subsp. *lactis* (*L. lactis*)
  - *L. delbrueckii* subsp. *bulgaricus* (*Lactobacillus bulgaricus*)
- *Leuconostoc*.

Coliforms: coliforms are facultative anaerobes with an optimum growth at 37° C. Coliforms are indicator organisms; they are closely associated with the presence of pathogens but not necessarily pathogenic themselves.

They also can cause rapid spoilage of milk because they are able to ferment lactose with the production of acid and gas, and are able to degrade milk proteins. They are killed by HTST treatment, therefore, their presence after treatment is indicative of contamination. *Escherichia coli* is an example belonging to this group.

Significance of microorganisms in milk:
- Information on the microbial content of milk can be used to judge its sanitary quality and the conditions of production
- If permitted to multiply, bacteria in milk can cause spoilage of the product
- Milk is potentially susceptible to contamination with pathogenic microorganisms. Precautions must be taken to minimize this possibility and to destroy pathogens that may gain entrance
- Certain microorganisms produce chemical changes that are desirable in the production of dairy products such as cheese, yogurt.

Spoilage Microorganisms in Milk

The microbial quality of raw milk is crucial for the production of quality dairy foods. Spoilage is a term used to describe the deterioration of a foods' texture, colour, odour or flavour to the point where it is unappetizing or unsuitable for human consumption. Microbial spoilage of food often involves the degradation of protein, carbohydrates, and fats by the microorganisms or their enzymes.

In milk, the microorganisms that are principally involved in spoilage are psychrotrophic organisms. Most psychrotrophs are destroyed by pasteurization temperatures, however, some like *Pseudomonas fluorescens, Pseudomonas fragi* can produce proteolytic and lipolytic extracellular enzymes which are heat stable and capable of causing spoilage.

Some species and strains of *Bacillus, Clostridium, Cornebacterium, Arthrobacter, Lactobacillus, Microbacterium, Micrococcus* , and *Streptococcus* can survive pasteurization and grow at refrigeration temperatures which can cause spoilage problems.

Pathogenic Microorganisms in Milk

Hygienic milk production practices, proper handling and storage of milk, and mandatory pasteurization has decreased the threat of milkborne diseases such as tuberculosis, brucellosis, and typhoid fever. There have been a number of foodborne illnesses resulting from the ingestion of raw milk, or dairy products made with milk that was not properly pasteurized or was poorly handled causing post-processing contamination. The following bacterial pathogens are still of concern today in raw milk and other dairy products:

- *Bacillus cereus*
- *Listeria monocytogenes*
- *Yersinia enterocolitica*
- *Salmonella* spp.
- *Escherichia coli* O157:H7
- *Campylobacter jejuni* .

It should also be noted that moulds, mainly of species of *Aspergillus* , *Fusarium* , and *Penicillium* can grow in milk and dairy products. If the conditions permit, these moulds may produce mycotoxins which can be a health hazard.

HACCP

Raw and end-products may be tested for the presence, level, or absence of microorganisms. Traditionally these practices were used to reduce manufacturing defects in dairy products and ensure compliance with specifications and regulations, however, they have many drawbacks:

1. destructive and time consuming
2. slow response

3. small sample size
4. delays in the release of the food.

In the 1960's, the Pillsbury Company, the U.S. Army, and NASA introduced a system for assuring pathogen-free foods for the space program. This system, called Hazard Analysis and Critical Control Points (HACCP), is a focus on critical food safety areas as part of total quality programs. It involves a critical examination of the entire food manufacturing process to determine every step where there is a possibility of physical, chemical, or microbiological contamination of the food which would render it unsafe or unacceptable for human consumption. These identified points are the critical control points (CCP). There are seven prinicples to HACCP:

1. analyze hazards
2. determine CCPs
3. establish critical limits
4. establish monitoring procedures
5. establish deviation procedures
6. establish verification procedures
7. establish record keeping procedures.

Before these principles can be put into place, a prerequisite program and preliminary setup is necessary.

Prerequisite Program:
- premise control
- receiving and storage control
- equipment performance and maintenance control
- personnel training
- sanitation
- recall procedure.

Preliminary Setup:
- assemble team
- describe the product
- identify intended use
- construct flow diagram and plant schematic
- verify the diagram on-site.

Food Safety Enhancement Program-FSEP is The Canadian Food Inspection Agency's HACCP initiative. There is extensive information at their Web site regarding FSEP, including implementation manuals, HACCP curriculum guidelines, and generic models.

Starter Cultures

Starter cultures are those microorganisms that are used in the production of cultured dairy products such as yogurt and cheese. The natural microflora of the milk is either inefficient, uncontrollable, and unpredictable, or is destroyed altogether by the heat treatments given to the milk. A starter culture can provide particular characteristics in a more controlled and predictable fermentation. The primary function of lactic starters is the production of lactic acid from lactose. Other functions of starter cultures may include the following:

- flavour, aroma, and alcohol production
- proteolytic and lipolytic activities
- inhibition of undesirable organisms.

There are two groups of lactic starter cultures:

1. simple or defined: single strain, or more than one in which the number is known
2. mixed or compound: more than one strain each providing its own specific characteristics.

Starter cultures may be categorized as mesophilic or thermophilic:

Mesophilic

- *Lactococcus lactis* subsp. *cremoris*
- *L. delbrueckii* subsp. *lactis*
- *L. lactis* subsp. *lactis* biovar *diacetylactis*
- *Leuconostoc mesenteroides* subsp. *cremoris*.

Thermophilic

- *Streptococcus salivarius* subsp. *thermophilus* (*S.thermophilus*)
- *Lactobacillus delbrueckii* subsp. *bulgaricus*
- *L. delbrueckii* subsp. *lactis*
- *L. casei*
- *L. helveticus*
- *L. plantarum* .

Mixtures of mesophilic and thermophilic microorganisms can also be used as in the production of some cheeses.

Bacteriophage

Bacteriophages are viruses that require bacteria host cells for growth and reproduction. Initially, the bacteriophage attaches itself to the bacteria cell wall and injects nuclear substance into the cell. Inside the cell, the nuclear substance produces shells, or phage coats, for the new bacteriophage which are quickly filled with nucleic acid. The bacterial cell ruptures and dies as the new bacteriophage are released. Bacteriophages are ubiquitous but generally enter the milk processing plant with the farm milk. They can be inactivated heat treatments of 30 min at 63 to 88° C, or by the use of chemical disinfectants.

Bacteriophages are of most concern in cheese making. They attack and destroy most of the lactic acid bacteria which prevents normal ripening known as slow or dead vat.

Starter Culture Preparation

Commercial manufacturers provide starter cultures in lyophilized (freeze-dryed), frozen or spray-dried forms. The dairy product manufacturers need to inoculate the culture into milk or other suitable substrate. There are a number of steps necessary for the propagation of starter culture ready for production:

1. Commercial culture
2. Mother culture — first inoculation; all cultures will originate from this preparation
3. Intermediate culture — in preparation of larger volumes of prepared starter
4. Bulk starter culture — this stage is used in dairy product production.

2

Controlling Microbial Quality of Dairy Products

Food quality is the quality characteristics of food that is acceptable to consumers. This includes external factors as appearance (size, shape, colour, gloss, and consistency), texture, and flavour; factors such as federal grade standards (e.g. of eggs) and internal (chemical, physical, microbial).

Food quality is enforced by the Food Safety Act 1990. Members of the public complain to trading standards professionals, who submit complaint samples and also samples used to routinely monitor the food marketplace to Public Analysts. Public Analysts carry out scientific analysis on the samples to determine whether the quality is of sufficient standard.

Food quality is an important food manufacturing requirement, because food consumers are susceptible to any form of contamination that may occur during the manufacturing process. Many consumers also rely on manufacturing and processing standards, particularly to know what ingredients are present, due to dietary, nutritional requirements (kosher, halal, vegetarian), or medical conditions (e.g., diabetes, or allergies).

Besides ingredient quality, there are also sanitation requirements. It is important to ensure that the food processing environment is as clean as possible in order to produce the safest possible food for the consumer. A recent example of poor sanitation recently has been the 2006 North American E. coli outbreak involving spinach, an outbreak that is still under investigation after new information has come to light regarding the involvement of Cambodian nationals.

Food quality also deals with product traceability, e.g. of ingredient and packaging suppliers, should a recall of the product be required. It also deals with labeling issues to ensure there is correct ingredient and nutritional information.

Recommended Minimum Internal Quality Control in Food Microbiology Testing Laboratories

National Standard Methods, which include standard operating procedures (SOPs), algorithms and guidance notes, promote high quality practices and help to assure the comparability of diagnostic information obtained in different laboratories. This in turn facilitates standardisation of surveillance underpinned by research, development and audit and promotes public health and patient confidence in their healthcare services.

The methods are well referenced and represent a good minimum standard for clinical and public health microbiology. However, in using National Standard Methods, laboratories should take account of local requirements and may need to undertake additional investigations.

The methods also provide a reference point for method development. National Standard Methods are developed, reviewed and updated through an open and wide consultation process where the views of all participants are considered and the resulting documents reflect the majority agreement of contributors.

Representatives of several professional organisations, including those whose logos appear on the front cover, are members of the working groups which develop National Standard Methods. Inclusion of an organisation's logo on the front cover implies support for the objectives and process of preparing standard methods. The representatives participate in the development of the National Standard Methods but their views are not necessarily those of the entire organisation of which they are a member.

The performance of standard methods depends on the quality of reagents, equipment, commercial and in-house test procedures. Laboratories should ensure that these have been validated and shown to be fit for purpose. Internal and external quality assurance procedures should also be in place.

Whereas every care has been taken in the preparation of this publication, the Health Protection Agency or any supporting organisation cannot be responsible for the accuracy of any statement or representation made or the consequences arising from the use of or alteration to any information contained in it.

These procedures are intended solely as a general resource for practicing professionals in the field, operating in the UK, and specialist advice should be obtained where necessary. If you make any changes to this publication, it must be made clear where changes have been made to the original document. The Health Protection Agency (HPA) should at all times be acknowledged.

The HPA is an independent organisation dedicated to protecting people's health. It brings together the expertise formerly in a number of official organisations.

The HPA aims to be a fully Caldicott compliant organisation. It seeks to take every possible precaution to prevent unauthorised disclosure of patient details and to ensure that patient-related records are kept under secure conditions. This Guidance Note makes recommendations for the minimum internal quality control (IQC) that food microbiology testing laboratories should implement to meet UKAS requirements.

Accreditation Scheme Requirements

Accreditation schemes stipulate some requirements for IQC. UKAS accreditation requires laboratories to have quality control procedures for monitoring the validity of tests undertaken.

General Requirements

Each member of staff should aim to process at least one IQC sample per year for each UKAS accredited test for which they are trained. This is to demonstrate ongoing competency and to demonstrate that each method is giving satisfactory performance. Staff who are not trained in an entire method should process the parts in which they are competent.

Laboratories should also aim to perform IQC tests at a frequency related to their level of sample testing.

Results should be recorded on worksheets. Copies of the worksheets may be stored in the individual members of staff training records or reference made to them to provide traceability.

All unexpected results should be investigated and corrective action taken where required. Investigations and actions should be documented.

Although this Guidance Note recommends preparing suspensions of organisms from reference cultures an alternative to this is to use ready prepared materials deemed suitable for the purpose. Such samples should provide a standardised inoculum at any pre-determined count for the required organisms.

EQA samples can also be used to demonstrate competence and method performance.

Preparation of Serial Dilutions

Serial dilutions are prepared as follows:
- Inoculate 1ml of an overnight broth culture (approximately 10^9 cfu/ml) of the target organism into 9ml of a suitable diluent (eg maximum recovery diluent) to give a 10-1 dilution. Using a separate pipette for each dilution, prepare serial 10-fold dilutions to 10-7.
- Perform a colony count on the serial dilution to calculate the actual cfu/ml of organisms.

Preparation of Spiked Samples

Spiked samples are prepared as follows:
1. To prepare a spiked sample containing approximately 10^2 cfu/g add 2ml of the 10-6 d dilution to 225ml of food suspension (225ml of diluent containing 25g of food).
2. To prepare a spiked sample containing approximately 10^4 cfu/g add 2ml of the 10-4 dilution to 225ml of food suspension (225ml of diluent containing 25g of food).
3. To prepare a spiked sample containing very low numbers (10-100) of cfu/25g add 1ml of the 10-7 dilution to 225ml of food suspension (225ml of diluent containing 25g of food). For detection methods the food is suspended in preenrichment or enrichment media.

Note: any food matrix used to prepare spiked samples should be sterile or have a very low background count or be negative for the target organism.

Note: Preparation of spiked food samples should not be done in the routine food laboratory.

Duplicate Testing

Duplicate spiked or real samples should be set up and tested by different staff if possible. Processing duplicate samples using different staff allows variation between staff to be monitored as well as between different portions of the same sample.

One set of results from duplicate samples should be designated in advance to be used for reporting. The other should be used for comparison only. An exception to this may be where duplicate testing forms part of the test protocol agreed with the customer, in which case the mean result should be used. If the results differ then the designated senior member of staff should be notified.

Duplicate counts from enumeration procedures should fall within 0.5 log10.

Replicate Testing

Guidance on replicate testing used to estimate the uncertainty of testing is documented.

Detection Methods

Spiked samples for detection methods are processed as follows:
1. For detection methods such as *Campylobacter, Escherichia coli* O157,*Listeria, Salmonella* and *Vibrio* species.
   — Prepare a spiked sample containing very low numbers of cfu/g of one of the following organisms:
     - *Campylobacter jejuni* NCTC 11322
     - *Escherichia coli* O157 NCTC 12900
     - *Listeria monocytogenes* NCTC 11994
     - *Salmonella poona* NCTC 4840
     - *Vibrio parahaemolyticus* NCTC 10885.

Note: If a verocytotoxin producing strain of *E.coli* is used then cultures must be handled in containment level 3.

Process according to the specific test method(s) for the target organism.

For detection methods, isolation of the target organism indicates that the method has worked satisfactorily. If the organism has not been isolated then the process should be repeated using two or more spiked samples

Enumeration Methods

Spiked samples for enumeration methods are processed as follows:
1. For enumeration methods such as aerobic colony count, coliforms, *Escherichia coli, Listeria* species, and *Staphylococcus aureus*.

- Prepare a spiked sample containing 10^2 cfu/g of each of the following organisms:

E. coli NCTC 9001
- *L. monocytogenes* NCTC 11994
- *S. aureus* NCTC 6571.

Process according to the following specified test method(s):
- Aerobic colony count at 30oC/48h
- Enumeration of coliforms/*E. coli*
- Enumeration of Enterobacteriaceae
- Enumeration of *Listeria* species
- Enumeration of *S. aureus*.

For enumeration methods such as *Bacillus cereus* and *Clostridium perfringens*.

Prepare a spiked sample containing 10^4 cfu/g of each of the following organisms:
- *B. cereus* NCTC 7464
- *C. perfringens* NCTC 8237.

Process according to the following specified test method(s):
- Colony count at 30oC/48h
- Enumeration of *B. cereus* and other *Bacillus* species
- Enumeration of *C. perfringens*.

For enumeration methods such as yeasts and moulds.

Prepare a spiked sample containing 10^2 cfu/g of each of the following organisms:
- *Saccharomyces cerevisiae* NCPF 3178
- *Aspergillus niger* NCPF 2275.

Process according to the following specified test method:
- Enumeration of yeasts and moulds.

For enumeration methods, counts should be compared with those expected to ensure adequate recovery and check that calculations are correct.

Laboratories should have a range for acceptable counts (e.g. within 0.5 log of expected count) that allows for a decision to be made as to whether the test has passed or failed.

An Overview of the Category Framework for Assessing Raw Milk Products

Food Standards Australia New Zealand (FSANZ) is currently assessing the requirements in the Australia New Zealand Food Standards Code for the sale of raw milk products in Australia through Proposal P1007 Primary Production and Processing Requirements for Raw Milk Products. As part of Proposal P1007, FSANZ has developed a framework in which to consider the various products that could be considered. This Category Framework Approach defines three categories, taking into account the effect of production methods and the intrinsic characteristic of the final product in assessing the potential food safety risk.

FSANZ has based the development of the Category Framework approach on preliminary results of microbiological risk assessments currently being undertaken by the agency, which have examined production factors and intrinsic properties of selected dairy products. Processing factors include curd cooking temperature, acidification and storage time. Intrinsic factors include moisture content, acidity and salt concentrations.

Category Framework Approach

The three categories in the framework are defined by the effect that production methods and intrinsic characteristics of the final products have on pathogen survival and growth. If the survival and growth of pathogens is more likely in some products, these products present a greater food safety risk compared to products where pathogen growth and survival is less likely.

While cheese is the major commercial raw milk product being considered, the framework approach will endeavour to achieve outcomes that are applicable to all products, including cultured milk, yoghurt, butter, ice cream etc. Therefore, the scope of this Proposal will examine all activities associated with the production of raw milk products, from on-farm milk production through to retail sale and use by the consumer.

Category 1

Products in this category are defined as those products where intrinsic characteristics and/or processing techniques eliminatepathogens that may have been present in the raw milk.

Examples of products in this category would include the extra hard grating cheeses. These cheeses are made from raw milk by heating the curd to greater than 48 ° C, have low moisture content (<36%) and a long maturation/ripening period – steps that result in the death of pathogens and that lead to cheeses have an equivalent level of safety to pasteurised products.

Category 2

Products in this category are defined as those products where intrinsic characteristics and/or processing techniques may allow the survival of pathogens present in the raw milk, but do not support the growth of these pathogens.

This category would apply to products where there is survival, but not growth, of pathogens that may have been present in the raw milk. In this case, control measures would need to ensure that the raw milk used to produce products is of an appropriate microbiological quality and that processing steps would achieve the necessary critical limits to control pathogens.

To produce these products, potentially a combination of control measures and verification activities will need to be applied to provide an acceptable level of microbial safety for consumption by the general population.

Category 3

Products in this category are defined as those products where intrinsic characteristics and/or processing conditions are likely to allow the survival of pathogens that may have been present in the raw milk and may support the growth of these pathogens.

Category 3 products are those where the processing steps that are applied would not reduce pathogens to an acceptable level. In general, if pathogens are present, they would be expected to multiply during manufacture. This category would include products such as raw drinking milk including raw goat's milk and likely to include high-moisture content cheeses.

Application of Category Framework Approach

Following the completion of the risk assessment and consumer research, FSANZ will use the outcomes to assess whether the nature of the hazards and the level of risk warrant permissions within the categories for the production of raw milk products for sale.

This would involve a variation to current regulatory requirements for the production of milk and dairy products in the Food Standards Code and other industry management interventions.

The appropriate control measures underpinning any permission will be presented in assessments reports prepared during the development of a Primary Production and Processing Standard for Raw Milk Products.

Bacteriological Tests for Milk

Milk testing and quality control is an essential component of any milk processing industry whether small, medium or large scale. Milk being made up of 87% water is prone to adulteration by unscrupulous middlemen and unfaithful farm workers. Moreover, its high nutritive value makes it an ideal medium for the rapid multiplication of bacteria, particularly under unhygienic production and storage at ambient temperatures. We know that, in order for any processor to make good dairy products, good quality raw materials are essential. A milk processor or handler will only be assured of the quality of raw milk if certain basic quality tests are carried out at various stages of transportation of milk from the producer to the processor and finally to the consumer.

There are a number of standard manuals and text books on milk quality control. However these may not be easily available to the emerging small scale to medium scale processors in Kenya. For these reasons, the Training Programme for the Small Scale Dairy Sector under project GOK/FAO/TCP/KEN/6611, has assembled this guide on Milk Testing and Quality Control so that it may be used for training and by the private small scale dairy processors. The methods selected are simple and basic and will suffice the requirements of most milk quality control laboratories of small scale processing units. For the larger plants with bigger laboratories more tests are to be found in the bibliography at the end of this booklet.

Milk Testing and Quality Control

What is Milk Quality Control?

Milk quality control is the use of approved tests to ensure the application of approved practices, standards and regulations concerning the milk and milk products. The tests are designed to ensure that milk products meet accepted standards for chemical composition and purity as well as levels of different micro-organisms.

Controlling Microbial Quality of Dairy Products

Why have Milk Quality Control?

Testing milk and milk products for quality and monitoring that milk products, processors and marketing agencies adhere to accepted codes of practices costs money. There must be good reasons why we have to have a quality control system for the dairy industry in Kenya.

The reasons are:

To the Milk Producer : The milk producer expects a fair price in accordance with the quality of milk she/he produces.

The Milk Processor : The milk processor who pays the producer must assure himself/herself that the milk received for processing is of normal composition and is suitable for processing into various dairy products.

The Consumer : The consumer expects to pay a fair price for milk and milk products of acceptable to excellent quality.

The Public and Government Agencies: These have to ensure that the health and nutritional status of the people is protected from consumption of contaminated and sub-standard foodstuffs and that prices paid are fair to the milk producers, the milk processor and the final consumer.

All the above-is only possible through institution of a workable quality testing and assurance system conforms to national or internationally acceptable standards.

Quality Control in the Milk Marketing Chain in Kenya

At the farm : Quality control and assurance must begin at the farm. This is achieved through farmers using approved practices of milk production and handling; and observation of laid down regulations regarding, use of veterinary drugs on lactating animals, regulations against adulterations of milk etc.

At Milk collection Centres : All milk from different farmers or bulked milk from various collecting centres must be checked for wholesomeness, bacteriological, and chemical quality.

At the Dairy Factories : Milk from individual farmers or bulked milk from various collecting centres

Within the Dairy Factory : Once the dairy factor has accepted the farmer milk it has the responsibility of ensuring that the milk is handled hygienically during processing. It must carry out quality assurance test to ensure that the products produced conform to specified standards as to the adequacy of effect of processes applied and the keeping quality of manufactured products. A good example is the phosphatase test used on pasteurised milk and the acidity development test done on U.H.T milk.

During marketing of processed products : Public Health authorities are employed by law to check the quality of food stuffs sold for public consumption and may impound substandard or contaminated foodstuffs including possible prosecution of culprits. This is done in order to protect the interest of the milk consuming public.

Techniques Used in Milk Testing and Quality Control

Milk Sampling

Accurate sampling is the first pre-requisite for fair and just quality control system. Liquid milk in cans and bulk tanks should be thoroughly mixed to disperse the milk fat before a milk sample is taken for any chemical control tests. Representative samples of packed products must be taken for any investigation on quality. Plungers and dippers me used in sampling milk from milk cans.

Sampling Milk for Bacteriological Testing

Sampling milk for bacteriological tests require a lot of care. Dippers used must have been sterilised in an autoclave or pressure cooker for at least 15mm at 120° C before hand in order not to contaminate the sample. On the spot sterilisation may be employed using 70% Alcohol swab and flaming or scaling in hot steam or boiling water for 1 minute.

Preservation of Sample

Milk samples for chemical tests : Milk samples for butterfat testing may be preserved with chemicals like Potassium dichromate. Milk samples that have been kept cooling a refrigerator or ice-box must first be warmed in water bath at 40 °C, cooled to 20°C, mixed and a sample then taken for butterfat determination. Other preservative chemicals include Sodium azid at the rate of 0.08% and Bronopol (2-bromo-2-nitro-1,3-propanediol) used at the rate of 0.02%.

If the laboratory cannot start work on a sample immediately after sampling, the sample must be cooled to near freezing point quickly and be kept cool till the work can start. If samples are to be taken in the field e.g. at a milk cooling centre, ice boxes with ice pecks are useful.

Labelling and Records Keeping

Samples must be clearly labelled with name of farmer or code number and records of dates, and places included in standard data sheets. Good records must be kept neat and in a dry place. It is desirable that milk producers should see their milk being tested, and the records should be made available to them if they so require.

Common Testing of Milk

Organoleptic Tests

The organoleptic test permits rapid segregation of poor quality milk at the milk receiving platform. No equipment is required, but the milk grader must have good sense of sight, smell and taste. The result of the test is obtained instantly, and the cost of the test are low. Milk which cannot be adequately judged organoleptically must be subjected to other more sensitive and objective tests.

Procedure:
- Open a can of milk.
- Immediately smell the milk.
- Observe the appearance of the milk.
- If still unable to make a clear judgement, taste the milk, but do not swallow it. Spit the milk sample into a bucket provided for that purpose or into a drain basin, flush with water.
- Look at the can lid and the milk can to check cleanliness.

Judgement

Abnormal smell and taste may be caused by:
- Atmospheric taint (e.g. barny/cowy odour).
- Physiological taints (hormonal imbalance, cows in late lactation-spontaneous rancidity).
- Bacterial taints.
- Chemical taints or discolouring.
- Advanced acidification (pH < 6.4).

Clot on Boiling (C.O.B) Test

The test is quick and simple. It is one of the old tests for too acid milk (pH<5.8) or abnormal milk (e.g. colostral or mastitis milk). If a milk sample fails in the test, the milk must contain many acid or rennet producing microorganisms or the milk has an abnormal high percentage of proteins like colostral milk. Such milk cannot stand the heat treatment in milk processing and must therefore be rejected.

Procedure : Boil a small amount of milk in a spoon, test tube or other suitable container. If there is clotting, coagulation or precipitation, the milk has failed the test. Heavy contamination in freshly drawn milk cannot be detected, when the acidity is below 0.20-0.26% Lactic acid.

The Alcohol Test

The test is quick and simple. It is based on instability of the proteins when the levels of acid and/or rennet are increased and acted upon by the alcohol. Also increased levels of albumen (colostrum milk) and salt concentrates (mastitis) results in a positive test.

Procedure : The test is done by mixing equal amounts of milk and 68% of ethanol solution in a small bottle or test tube. (68 % Ethanol solution is prepared from 68 mls 96% (absolute) alcohol and 28 mls distilled water). If the tested milk is of good quality, there will be no coagulation, clotting or precipitation, but it is necessary to look for small lumps. The first clotting due to acid development can first be seen at 0.21-0.23% Lactic acid. For routine testing 2 mls milk is mixed with 2 mls 68% alcohol.

The Alcohol-Alizarin Test

The procedure for carrying out the test is the same as for alcohol test but this test is more informative. Alizarin is a colour indicator changing colour according to the acidity. The Alcohol Alizarin solution can be bought ready made or be prepared by adding 0.4 grammes alizarin powder to 1 litre of 61% alcohol solution.

Acidity Test

Bacteria that normally develop in raw milk produce more or less of lactic acid. In the acidity test the acid is neutralised with 0.1 N Sodium hydroxide and the amount of alkaline is measured. From this, the percentage of lactic acid can be calculated. Fresh milk contains in this test also "natural acidity" which is due to the natural ability to resist pH changes .The natural acidity of milk is 0.16 - 0.18%. Figures higher than this signifies developed acidity due to the action of bacteria on milk sugar.

Apparatus:
- A porcelain dish or small conical flask
- 10 ml pipette, graduated

- 1 ml pipette
- A Burette, 0.1 ml graduations
- A glass rod for stirring the milk in the dish
- A Phenophtalein indicator solution, 0.5% in 50% Alcohol
- N Sodium hydroxide solution.

Procedure : 9 ml of the milk measured into the porcelain dish/ conical flask, 1 ml Phenopthalein is added and then slowly from the burret, 0.1 N Sodium hydroxide under continuous mixing, until a faint pink colour appears.

The number of mls of Sodium hydroxide solution divided by 10 expresses the percentage of lactic acid.

Resazurin Test

Resazurin test is the most widely used test for hygiene and the potential keeping quality of raw milk. Resazurin is a dye indicator. Under specified conditions Resazurin is dissolved in distilled boiled water. The Resazurin solution can later be used to test the microbial activity in a given milk sample.

Resazurin can be carried out as:
i. 10 min test.
ii. 1 hr test.
iii. 3 hr test.

The 10 min Resazurin test is useful and rapid, screening test used at the milk platform.

The 1 hr test and 3 hr tests provide more accurate information about the milk quality, but after a fairy long time. They are usually carried out in the laboratory.

Apparatus and reagents:
- Resazurin tablets
- Test tubes with 10 mls mark
- 1 ml pipette or dispenser for Resazurin solution.
- Water bath thermostatically controlled
- Lovibond comparator with Resazurin disc 4/9.

Procedure : The solution of Resazurin as prepared by adding one tablet to 50 mls of distilled sterile water. Rasazurin solution must not be exposed to sunlight, and it should not be used for more than eight hours because it losses strength.

Mix the milk and with a sanitized dipper put 10 mls milk into a sterile test tube. Add one ml of Resazurin solution, stopper with a sterile stopper, mix gently the dye into the milk and mark the tube before the incubation in a water bath, place the test tube in a Lovibond comparator with Resazurin disk and compare it colourimetrically with a test tube containing 10 ml milk of the same sample, but without the dye (Blank).

Table 1: Readings and Results (10 Minute Resazurin Test)

Resazurin disc No.	Colour	Grade of milk	Action
6	Blue	Excellent	Accept
5	Light blue	v. good	Accept
4	Purple	Good	Accept
3	Purple pink	Fair	Separate
2	Light pink	Poor	Separate
1	Pink	Bad	Reject
0	white	Very bad	Reject

The Gerber Butterfat Test

The fat content of milk and cream is the most important single factor in determining the price to be paid for milk supplied by farmers in many countries. Also, in order to calculate the correct amount of feed ration for high yielding dairy cows, it is important to know the butterfat percentage as well as well as the yield of the milk produced. Further more the butterfat percentage in the milk of individual animals must be known in many breeding programmes.

Butterfat tests are also done on milk and milk products in order to make accurate adjustments of the butterfat percentage in standardised milk and milk products.

Apparatus for DF test:
- Gerber butyrameters, 0-6% or 0-8% Of
- Rubber stoppers for butyrometers
- 10.94 or 11 ml pipettes for milk
- 10 mls pippetes or dispensers for Gerber Acid
- 1 mls pippetes or dispensers for Amyl alcohol
- stands for butyrometers.

Gerber water bath Reagents:
- Gerber sulphuric acid, (1.82 g/cc)
- Amyl alcohol.

Treatment of Samples

Fresh milk at approximately 20°C should be mixed well. Samples kept cool for some days should be warmed to 40°C, mixed gently and cooled to 20°C before the testing.

Procedure: Add 10 mls sulphuric acid to the butyrometer followed by 10.94 or 11 mls of well mixed milk. Avoid wetting of the neck of the butyrometer.

Next add 1 ml of Amyl alcohol, insert stopper and shake the butyrometer carefully until the curd dissolves and no white particles can be seen. Place the butyrometer in the water bath at 65°C and keep it there until a set is ready for centrifuging. The butyrometer must be placed in the centrifuge with the stem (scale) pointing towards the centre of the centrifuge.

Spin for 5 min. at 1100 rpm.

Remove the butyrometers from the centrifuge.

Put the butyrometers in a water bath maintained at 65°C for 3 min. before taking the reading.

(Note: When transferring the butyrometers from the centrifuge into the water bath make sure that the butyrometers are all the time held with the neck pointing up).

The fat column should be read from the lowest point of the meniscus of the interface of the acid-fat to the 0-mark of the scale and read the butterfat percentage.

The butyrometers should be emptied into a special container for the very corrosive liquid of acid-milk, and the butyrometers should be washed in warm water and dried before the next use.

Appearance of the Test

The colour of the fat column should be straw yellow.

The ends of the fat column should be clearly and sharply defined.

The fat column should be free from specks and sediment.

The water just below the fat column should be perfectly clear.

The fat should be within the graduation.

Problems in Test Results

Curdy tests:
- Too lightly coloured or curdy fat column can be due to:
- Temperature at milk or acid or both too low.

- Acid too weak.
- Insufficient acid.
- Milk and acid not mixed thoroughly.

Charred tests:
- Darkened fat column containing black speck at the base is due to:
- Temperature of milk-acid mixture too high.
- Acid too strong.
- Milk and acid mixed too slowly.
- Too much acid used.
- Acid dropped through the milk.

The Lactometer Test

Addition of water to milk can be a big problem where we have unfaithful farm workers, milk transporters and greedy milk hawkers. A few farmers may also fall victim of this illegal practice. Any buyer of milk should therefore assure himself/herself that the milk he/she purchases is wholesome and has not been adulterated. Milk has a specific gravity.

When its adultered with water or other materials are added or both misdeeds are committed, the density of milk change from its normal value to abnormal. The lactometer test is designed to detect the change in density of such adulterated milk. Carried out together with the Gerber butterfat test, it enables the milk processor to calculate the milk total solids (% TS) and solids not fat (SNF). In normal milk SNF should not be below 8.5% according to Kenya Standards (KBS No 05-l0:-1976).

Procedure: Mix the milk sample gently and pour it gently into a measuring cylinder (300-500). Let the Lactometer sink slowly into the milk. Read and record the last Lactometer degree (°L) just above the surface of the milk. If the temperature of the milk is different from the calibration temperature (Calibration temperature may be=20 0C) of the lactometer, calculate the temperature correction. For each °C above the calibration temperature add 0.2°L; for each °C below calibration temperature subtract 0.2 °L from the recorded lactometer reading.

Table 2: Example Calibration temperature of lactometer 20°C

Sample	Milk temperature	Lactometer reading	Correction	True reading
No.1	17 °C	30.6 °L	- 0.6 °L	30.0 °L
No.2	20 °C	30.0 °L	Nil	30.0 °L
No.3	23 °C	29.4 °L	+ 0.6 °L	30.0 °L

For the calculations, use lactometer degrees, and for the conversion to density write 1.0 in front of the true lactometer reading ,i.e. 1.030 g/ml. Clever people may try to adulterate milk in such a way that the lactometer cannot show the adulteration. But look to see if there is an unusual sediment from the milk at the bottom of the milk can and taste to find out if the milk is too sweet or salty to be normal. Samples of milk from individual cows often have lactometer reading outside the range of average milk, while samples of milk from herds should have readings hear the average milk, but wrong feeding, may result in low readings. Kenyan standards expects milk to have specific gravity of 1.026 -1.032 g/ml which implies a Lactometer reading range of 26.0 -32.0 °L. If the reading is consistently lower than expected and the milk supplier disputes any wrong doing arrange to take a genuine sample from the supplier (i.e. inspect milk right from source).

Freezing Point Determination

The freezing point of milk is regarded to be the most constant of all measurable properties of milk. A small adulteration of milk with water will cause a detectable elevation of the freezing point of milk from its normal values of -0.54°C. Since the test is accurate and sensitive to added water in milk, it is used to detect whether milk is of normal composition and adulterated.

Inhibitor Test

Milk collected from producers may contain drugs and/or pesticides residues. These when present in significant amounts in milk may inhibit the growth of lactic acid bacteria used in the manufacture of fermented milk such as Mala, cheese and Yoghurt, besides being a health hazard.

Principle of the method: The suspected milk sample is subjected to a fermentation test with starter culture and the acidity checked after three (3) hours. The values of the titratable acidity obtained is compared with titratable acidity of a similarly treated sample which is free from any inhibitory substances.

Materials:
- test tubes
- Starter culture
- lml pipette
- water bath
- material for determination of titratable acidity.

Procedure: Three test tubes are filled with 10 ml of sample to be tested and three test tubes filled with normal milk.

All tubes are heated to 90°C by putting them in boiling water for 3 - 5 minutes.

After cooling to optimum temperature of the starter culture (30,37, or 42°C), 1 ml of starter culture is added to each test tube, mixed and incubated for 3 hours.

After each hour, one test tube is from the test sample and the control sample is determined.

Assessment of results: If acid production in suspected sample is the same as the normal sample, then the suspect sample does not contain any inhibitory substances;

If acid production as suspect sample is less than in the normal milk sample, then, the suspect sample contains antibiotics or other inhibitory substances.

Quality Control of Pasteurised Milk

When milk is pasteurised at 63°C for 30 min in batch pasteuriser or 72°C for 15 seconds in heat exchanger, continuous flow pasteurisers, ALL PATHOGENIC BACTERIA ARE DESTROYED, there by rendering milk safe for human consumption. Simultaneously various enzymes present in milk, and which might affect its flavour, are destroyed.

In order to determine whether or not milk has been adequately pasteurised, one of the enzymes normally present in milk phosphatase, is measured. A negative phosphatase result indicates that the enzyme and any pathogenic bacteria have been destroyed during pasteursation. If it is positive, it means the pasteurisation process was inadequate and the milk may not be safe for human consumption and will have a short shelf life.

- Test tubes
- 5 mls pipettes

- 1 ml pipettes
- 100 ml volumetric flask
- 500 ml volumetric flask
- water bath at 37°C

Note: All glassware must be rinsed, cleaned, rinsed in chromic acid solution and boiled in water for 30 min.

Reagent

Buffer solution: Is mixed by 0.75g anhydrous sodium carbonate and 1.75g Sodium bicarbonate in 500 ml distilled water.

Buffer-substrate solution: Place 0.15 g of di-sodium paranitrophenylphosphate (the substrate) into a clean 100ml measuring cylinder.

Add the buffer solution to make to 100 ml mark.

Store this buffer-substrate solution in a refrigerator and protected against light. It should not be used after one week. Prepare a fresh stock.

Procedure: Pipette 5 mls buffer-substrate solution into a test tube, stopper and warm the solution in the water bath at 37° C. Add to the test tube 1ml of the milk to be tested, stopper and mix well and place in water bath at 37° C. Prepare a blank sample from boiled milk of the same type as that undergoing the test. Incubate both the test samples and the blank sample at 37° C. for 2 hrs. After incubation, remove the tubes and mix them thoroughly.

Place one sample against the blank in a Lovibond comparator" ALL PURPOSES" using A.P.T.W. disc and rotate the disc until the colour of the test sample is matched and read the disc number.

Interpretation

Disc Reading after 2 hrs incubation at 37° C.	Remarks
0-10	Properly pasteurised
10-18	Slightly under pasteurised
18-42	Under Pasteurised
> 42	Not Pasteurised.

3
Physico-chemical Testing of Milk and Dairy Products

Microbiological and physicochemical analysis of different UHT milks available in market Raw milk is milk in its natural (unpasteurized) state. Contaminated raw milk can be a source of harmful bacteria, such as those that cause undulant fever, dysentery, salmonellosis and tuberculosis. "Certified" milk, obtained from cows certified as healthy, is unpasteurized milk with a bacteria count below a specified standard, but it still can contain significant numbers of disease producing organisms.

Different heat and treatments are given to raw milk in order to remove pathogenic organisms, to increase the shelf life, to help subsequent processing e.g. for warming before separation and homogenization or as an essential treatment before cheese making, yoghurt manufacture and production of evaporated and dried milk products (Singh, 1993).

Pasteurization, sterilization (in bottle) and UHT (ultra-high-temperature) treatment integrated with aseptic packing. Sterilization (in bottle) is the term applied to a heat treatment process which has a bactericidal effect greater than pasteurization. Although it does not result in sterility, it gives the processed milk a longer shelf life. As a result of the long holding time at this elevated temperature, the product has a cooked flavour and a pronounced brown colour. Unlike sterilization, pasteurization is not intended to kill all pathogenic micro-organisms in the food or liquid. Instead, pasteurization aims to reduce the number of viable pathogens so they are unlikely to cause disease.

Ultra-high temperature (UHT or ultra-heat treated) is also used for milk treatment. UHT processing holds the milk at a temperature of 138°C (250°F) for a fraction of a second. Milk simply labelled "pasteurization" is usually treated with the HTST method, whereas milk labelled "ultra-pasteurization" or simply "UHT" has been treated with the UHT method (Bylund, 1995). Heating of milk accounts 2 main problems, age gelation and off flavour development, which limits shelf life of milk. UHT treatment of milk leads to a much larger production of small sized casein micelles compared to raw or pasteurized milk (Singh, 1993). Biochemical processes involve are heat resistance and reactivation of natural and bacterial proteases and survival of bacterial spores.

Proteolysis of UHT milk during storage at room temperature is a major factor limiting the shelf life through changes in its flavour and texture (Datta et al., 2002). The changes ultimately reduce the quality and limit the shelf life of UHT milk via development of off flavours, fat separation and sedimentation, which principally falls into 2 categories, liberation of volatile fatty acids such as butyric acid and oxidation of free or unsaturated fatty acids (Datta et al., 2002). Above 135°C the protein deposited on the fat globule membrane form a network which makes the membrane denser and less permeable.

There is an increase in acidity and viscosity with a decrease in pH with the storage time increased both in UHT. Clare et al. (2005) determined that sweet aromatic flavour and sweet taste of UHT milk decreases during storage.

The microorganisms, which cause spoilage in milk, which is intended to be sterile (UHT treatment), are either resistant types that have survived the heat treatment, or organisms that have contaminated the product after the sterilization process. Contamination may either be by heat labile organism or heat resistant forms such as spores. Contaminating spores are, however, likely to be less heat resistant than those, which might survive the heat treatment.

The problem of post treatment contamination of in container sterilized product is well known. The contamination can either through poor seal or through pinhole in the container. Post treatment contaminants in UHT milk may be either spores, which would not be expected to be heat resistant enough to survive the heat treatment or non heat resistant vegetative organisms. Organisms of first type will probably have entered from ineffectively sterilized plant down stream from the heat treatment stage of the process, which includes spores of *Bacillus cereus* and *Bacillus licheniformis*. Organisms of second type will probably have entered through poorly sealed container after aseptic filling.

The types of spores, which have been investigated as of particular relevance in the UHT, are those of *Bacillus stearothermophilus*, *Bacillus subtilis* and *Clostridium botulinum* has been studied. The high spore counts can occur at the dairy farm and that feed and milking equipment can act as reservoirs or entry points for potentially highly heat resistant spores into raw milk.

Lowering this spore load by good hygienic measures could probably further reduce the contamination level of raw milk, in this way minimizing the aerobic spore forming bacteria that could lead to spoilage of milk and dairy products (Westhoff and Dougerty, 1981). These problems had been reported internationally since long, hence the project was planned to observe the physicochemical. In this study the de-clared shelf life of different UHT milk available in market is studied.

Materials and Methods

The samples were taken in sterilized syringes for microbiological analysis and in clean stainless steel containers of 1 liter for chemical and sensory analysis. The samples were analyzed at interval of 1, 2, 3, 4, 5, 6, 7, 8, 9, 10, 11 and 12 weeks. During this period, samples were stored at room temperature (25°C) to provide them similar conditions, as they are stored in market. For microbiological analysis the samples were examined for aerobic plate counts (APC), *E. coli* counts, *B. cereus*, *B. subtilis* counts and for spore formers counts. The parameters examined for the chemical analysis were sedimentation, pH, and acidity as lactic acid %, fat % before and after shaking the milk, SNF % before and after shaking and protein % before and after shaking. For sensory evaluation colour, aroma and taste were examined.

Physicochemical Analysis of Milk

To assess the physical and chemical changes in processed milk samples following tests were carried out.

Sedimentation Test

Sedimentation test was performed by following modified method as described by Ramsey and Swartzel (1984). According to this method, milk was drain from the cartons leaving the bottom 4 cm.

The cartons were inverted for approximately 10 min, up righted and placed in the exhaust hood to dry. The cartons were allowed to dry for 48 h after the bottom flaps or wings of cartons had been opened to facilitate the drying of any sediment entrapped there. The cartons were weighted and then washed thoroughly to remove any sediment or residue adhering to the container. The washed cartons were again dried and weighted.

Solids Non Fats (SNF) %

Solids non fats (SNF) % was determined by lactometeric method as described by Ramsey and Swartzel (1984).

Total Titratable Acidity

Total titratable acidity determined according to the method of AOAC, (2005).

pH

The pH value of milk was determined by using a digital pH meter (AOAC, 2005). Prior to use, the pH meter was standardized with standard buffer solution of pH 4 and 7.

Fat

Milk fat % was determined by Gerber (1997) method as described by FAO (1997) by using the butyrometer.

Protein

The protein was estimated by formal titration method (Davide, 1977).

Microbial Analysis

Microbial analysis was performed according to standard methods (AOAC, 2005).

Total Viable Counts

The plate count agar media (Bridson, 1995) was used for the total viable count in UHT milk samples (AOAC, 2005). Plates were incubated for 24 h at 37°C.

Determination of Coliforms

Coliform counts were determined by pour plate method on violet red bile agar, prepared according to the manufacturer instructions. All plates were incubated at 37°C for 24 h.

Determination of Bacillus Species

B. cereus selective agar base (Bridson, 1995) is used for isolation and enumeration of *B. cereus* and *B. subtilis*. All plates were incubated at 37°C for 24 h.

Determination of Spore Formers

Plate count agar media (Bridson, 1995) is used for the enumeration of spore formers. Sterile medium was poured into sterile petri plates and allowed the medium to solidify. Sample is heated at 80°C for 10 min using water bath. These plates were inoculated with 1 ml sample by using sterile pipette. After inoculation, the sample was well mixed in the petri plates by to and fro motion. All plates were incubated in an inverted position for 72 h at 55°C.

E. coli Counts

For *E. coli* count MacConkey's agar (Bridson, 1995) was used. Sample from lactose positive tubes in case of coliform counts were applied directly on the MacConkey's agar (Bridson, 1995) plates and incubated at 37°C for 24 h.

Sensory analysis

The stored milk samples were evaluated sensorial for colour and flavor by scoring method as described by Larmond (1977).

Results and Discussion

The changes that have taken place during storage depend on temperature of storage, extent of exposure of the milk to light and availability of oxygen. The dairy company of Pakistan shows shelf life of 12 weeks on the labels of milk packs, during this mentioned period. Milk must be in best condition for consumption.

For storage time than a week or 2, these effects may be greater than those of the heat treatment. Changes in colour, flavour and texture are readily detected by the consumer and may reduce the acceptability of the products. Other changes cannot be recognized by the consumer and are not necessarily correlated with organoleptic, recognizable changes, but are of potential nutritional importance. The quality of sediment depends on the raw milk and on the type and severity of the heat treatments. For any 1 type of process, the amount of sediments increases in the severity of the heat treatment.

The amount of sediment decreases with homogenization pressure (Robinson, 1994). Results obtained from sedimentation test in UHT milk during storage period of 3 months (12 weeks) shows that there is an effect of heat processing and subsequent storage on sedimentation in all 4 samples of UHT milk. The changes started in week 2 of shelf life for samples I and III and sample II showed formation of sediments after week 6. Sample III reaches up to 7.10/250 gml-1, which is a considerable changes and sample II showed formation of sediments after week 5. The alcohol test can be used to detect raw milk that is likely to give high level of the normal type of sediments and there are indications that it may be useful in predicting the abnormal type (Sweetsur and White, 1975).

Processing operations influences acid base equilibrium in milk. UHT treatment results in a pH decrease, due to conversion of lactose into different organic acids. In milk, casein micelles are stable at natural pH, that is, 6.7. Lowering the pH facilitates aggregations of casein micelles and forms a gel. Results regarding effect of storage on pH of UHT processed milk during storage period of 90 days show that there is storage effect on pH level. Maximum pH value (6.81and 6.85 in samples 1 and 2, respectively and 6.75 in samples III and IV) was recorded in 1st week while minimum pH values obtained in 12th week of shelf life (6.20, 6.65, 6.17 and 6.55 in sample 1, 2, 3 and 4 of UHT milk respectively). Vankatachalm and McMahon (1991) verified drop in pH and they associated it with browning reactions. Andrews et al. (1977) confirmed similar effects and concluded that the level and extent of pH decrease was related to age gelation. When milk is heated at a temperature above 100oC and subsequent stored, lactose is degraded to acids. Formic acid is the principal acid produced due to which titratable acidity of milk rises.

Increase in free fatty acids is also responsible for increasing the total titratable acidity of milk (Swartzel, 1983). Results obtained by the analysis for total titratable acidity. The acidity value was 0.11% while during storage of UHT milk minimum acidity was recorded in 1st week and maximum value (0.18%) at 90 days life in sample 1 while 0.15 in case of sample 1 and 0.13 in samples 2 and 4. The proteins of milk are the constituents most affected by heating and subsequent storage of milk. The principal changes in UHT milk during storage may be due to enzymes. Many proteins in milk are very heat labile e.g. whey proteins, vitamin binding protein, antimicrobial proteins etc. These proteins coagulate after heating hence the texture of milk is deteriorated during storage (Fox and McSweeny, 1998).

Casein polymerization is greater at high storage temperature, but occurs significantly even under refrigeration: 50% of the protein may be in the polymer form after 6 months at 37°C, and 21% after 6 months at 4°C (Andrews, 1977). The results regarding protein % of stored UHT milk describes that there is effect of storage on protein contents of UHT processed milk. That is, in week 1 protein contents were 3.30%, 3.70% for sample 1 and 2 while in week 12 of storage were 2.35 and 3.48 respectively. In case of samples III and IV, protein contents were 3.40 in week 1 while it changes to 1.15 and 2.59, respectively, in week 12. There is no change in protein % in all samples after shaking of UHT milk. Chen et al. (2005) showed almost a 90% loss and denaturation of S-lactoglobulin (LG) of the UHT processed and dry milks by using polyacrylamide gel electrophoresis. Of the principle constituent, the fats are probably least affected by UHT treatment.

It is concluded from the whole study that there is an increase in sedimentation value, fat separation, titratable acidity during storage, while decrease was found in pH and protein % during storage of 12 weeks. The increase in sedimentation shows excessive protein denaturation during processing and subsequent storage. In UHT processed milks the fat separation was observed during storage.

This high % of fat separation is attributed with less homogenizing efficiency during processing. On microbiological examination, not any colony found on TPC plates, *coliform* agar plates, *E. coli* plate, *B. cereus, B. subtilus*, and spore formers plates, in all the 4 samples of UHT milk during storage of 12 weeks. Sensory characteristics showed a significant decrease in scores during storage. These all are factors that limit the shelf life of UHT milk.

The shelf life of milk mainly depend on the quality of raw milk and better quality of milk can be achieved in Pakistan, when the manufacturers have better milk collection system. The manufacturer of sample II has its own sophisticated type of milk collection system said to be VMCs (village milk collection centers). At these centers, milk is collected at small scale and in short time it is transported at low temperature to the processing plant, avoiding contamination, due to this practice the microbial as well as other contamination can be controlled in better way before heat treatment or processing. While manufacturers of other dairy industries of Pakistan get milk from contractors and ice added milk is mostly supplied to these industries which disturb the mineral balance and natural emulsion and give higher water activity which leads to physicochemical, microbiological as well as sensory changes during shelf life of milk.

Microbiological and Physicochemical Properties of Raw Milk Used for Processing Pasteurized Milk in Blue Nile Dairy Company (Sudan)

Milk is a highly nutritious food, ideal for microbial growth and the fresh milk easily deteriorates to become unsuitable for processing and human consumption (FAO 2001). High bacterial counts are indicator of poor production hygiene or ineffective pasteurization of milk (Harding 1999).

Milk and milk products derived from dairy cows milk can harbour a variety of microorganisms and can be important sources of foodborne pathogens. The presence of food-borne pathogens in milk is due to direct contact with contaminated sources in the dairy farm environment and to excretion from the udder of an infected animal (El Zubeir *et al.* 2006).

The hygienic quality problems of milk may arise from raw milk of diseased animals (Murphy and Boor 2000). Kang *et al.* (2005) reported that the presence of antimicrobial substances in raw milk could have serious toxicological and technical consequences. Raw milk may contain over 2,000,000 cfu/ml before processing of liquid milk or cheese making (Kameni *et al.* 2002). The raw milk distributed for consumption in Sudan does not find the real quality control measures needed to be of good quality food (Mohamed and El Zubeir 2007). However, some new private dairy plants started the processing of fluid milk and some dairy products. These are faced with many problems of which the quality control measures constitute an important concern. Hence, the present study was designed to assess the chemical, physical and microbial properties of raw milk supplied to the Blue Nile Dairy Company plant (CAPO) and to compare it with the produced pasteurized milk.

Materials and Methods

Source of Milk: This study was carried out during June to September 2005, and the raw milk for this study was collected from two dairies, namely Blue Nile Dairy Company and Kordi Farms. Blue Nile Dairy Company plant deals with both suppliers as a source of raw milk for processing pasteurized milk. Raw milk from both dairy farms is usually mixed before processing. Milk fat was standardized to 3-3.2%, and the milk was pasteurized at 72- 76< C for 15 second using a high temperature short time (HTST) plate heat exchanger (Wincantor Pasteurizer, Wincanton Engineering Ltd, South Street Sherborne Dorset, U.K.). The pasteurized milk was packed into Tetra Pack container (Tetra Pack Technical Services AB, Ruben Rausing gata, 5-221 86 Lund, Sweden).

Raw milk samples (36 samples) and pasteurized milk after processing and before packaging (12 samples) were examined for total bacterial counts (TBC), coliform, psychrotrophic (PC) and thermoduric bacterial counts. Physiochemical properties (fat, protein, lactose, ash, solid not fat, density, acidity, pH and freezing point), antibiotics and phosphatase test were also done.

Chemical Analysis of Milk Samples: The milk constituents (fat, protein, lactose, ash and solid not fat) and physical characteristics (density and freezing point), of the milk samples were determined by milk analyzer Lactoscan 90 (Aple Industries services–La Roche Sur Foron, France), according to manufacturer's instructions. Milk samples were mixed gently 4-5 times to avoid any air enclosure in the milk. Then 25 ml samples were taken in the sample-tube and put in the sample- holder one at a time with the analyzer in the recess position. Then when the starting button activated, the analyzer sucks the milk, makes the measurements, returns the milk in the sample-tube and the digital indicator (IED display) shows the specified results.

Antibiotic residues were determined using Delvotest® SP- ampule Kit (202– Delvotest SP 100, test box, DSM Food Specialties, the Netherlands). The method was carried out according to the manufacturer's instructions. Phosphatase test was done using Lactognost tables and powders obtained from Heyl, Chem. Pharm-Fabrik, 14167 Berlin. The procedure for phosphatase test was done according to the manufacturer's instructions. The acidity of the samples was determined according to AOAC (1990). The temperature and the pH of the samples were determined using pH– meter (Wagtech, HI 8314 membrane pH Meter, U.K.).

Microbiological Analysis: The samples were examined for TBC, coliforms, thermoduric and psychrotrophic counts according to Houghtby *et al.* (1992); Christen *et al.* (1992); Ballou *et al.* (1995); Ravanis and Lewis (1995), respectively. Plate count agar No. 298 (Biomark Laboratories) was used for enumeration of TBC, thermoduric bacteria and psychrotrophic counts, while violet red bile agar No. 779 (Biomark Laboratories) was used to determine coliform counts. The media were prepared according to manufacturer's instructions. Plates for enumeration of TBC, thermoduric bacteria and coliforms were incubated at 32° C for 48 hours, 37° C for 48 hours and 37° C for 24 hours, respectively. Plates for enumeration of psychrotrophic counts were incubated at 7° C for ten days. Developed colonies were counted using manual colony counter. The plates counting 25-250 colonies were selected as described by Houghtby et al. (1992). The number reciprocal of the dilution factor was recorded as colony forming unit per ml (cfu/ ml). The milk ring test (MRT) for brucellosis was carried out according to Harrigan and McCance (1976).

Statistical Analysis: Data were analyzed by SPSS programme (Statistical Package for Social Science, version 10.00). This test combines ANOVA with comparison of differences between means of the treatments at the significance level of $P < 0.05$.

Results and Discussion

The means of fat, protein, lactose, ash and solids not fat content were 4.14%, 3.48 %, 4.33%, 0.778% and 8.58% in raw milk samples mixture. The density, freezing point, titratable acidty and pH revealed 1.031, -0.520, 0.145 and 7.02. The analysis of variance showed highly significant variations ($P < 0.01$) due to the source of raw milk samples, except for fat. The composition of raw milks in the present study was compared favourably with the composition of milk in northern Europe, which contained fat of 4.3%, total protein of 3.4%, lactose of 4.65%, ash of 0.73%, TS of 13.3% and SNF of 9.0% (Invensys APV 2002). This result also agrees with that reported by El Zubeir et al. (2005) for raw milk. The present study revealed lower mean values for lactose (%) than that reported by El Zubeir et al. (2005). The lower lactose may be due to the effect of psychotrophic bacteria (Ballou et al. 1995). The results of physicochemical analysis of mixed raw milk used for producing fluid milk and the pasteurized milk. These results were higher compared to that reported by Elmagli and El Zubeir (2006a).

These differences in milk composition may be due to initial raw milk used and the procedure of processing. However the results of freezing point agreed with those reported by Tetra Pak Processing Systems (2003) for freezing points of raw and pasteurized milk -0.520± 0.001 and -0.447± 0.000, respectively obtained during the present study. This study also agreed with that reported by Elmagli and El

Zubeir (2006a) who found the freezing point was -0.4734± 0.05032 C. The obtained data for acidity o of raw and pasteurized milk of 0.145% and 0.143%, respectively, which are in line with that reported by Harding (1999), while the mean value was lower than that reported by Mohamed and El Zubeir (2007). The microbiological quality of the raw milk used for processing pasteurized milk showed that the initial quality was good for TBC (log 4.800 cfu\ml), coliform counts (log 4.157 cfu/ml), thermoduric bacterial counts (log 2.994) and psychotrophic bacterial counts (log 810 cfu/ml). The analysis of variance showed highly significant differences (P< 0.01) due to the source of raw milk samples for TBC and coliform counts. This result was lower than that reported by PMO (2001) for the average standard plate counts for can and bulk milk (700.000 bacteria /ml and 100.0 bacteria /ml, respectively). Moreover, the microbial standards for grade A raw milk is 100.0 bacteria/ml (PMO, 2001). The lower counts of bacteria may be due to good cleaning system and good handling from farms to the plant as was stated before by Chye et $al.$ (2004). Lower TBC value was obtained for pasteurized milk than that reported by Elmagli and El Zubeir (2006b), who reported a range of 6.5×105 to 6.5×1014, but was similar to that of Reena et $al.$ (2003). In addition, PMO (2001) reported that the bacteria standards for grade A pasteurized milk should be less than 20,000 bacteria /ml. Coliform bacteria counts of pasteurized milk showed lower numbers than these reported by Elmagli and El Zubeir (2006b). The lower coliform counts might be due to hygienic quality of raw milk, proper pasteurization process, good packaging and good storage conditions. This agreed with PMO (2001) who reported that the total bacterial standards for grade A pasteurized milk should be < 10 coliform/ ml. In addition, coliform counts obtained are in line with Sudanese Standards (SSMO, 2005) which stated that the maximum coliform counts should not to exceed 102 cfu/ml.

Thermoduric bacterial counts (log 0.621cfu/ml) was lower than that reported by Mohamed and El Zubeir (2007). However, the present findings agreed with that reported by Invensys APV (2002) who reported an aerobic spores-forming bacteria of <400. The mean value of psychrotrophic bacteria for pasteurized milk was log 0.360 cfu/ml, which was lower counts compared with that reported by Elmagli and El Zubeir (2006b), who found the psychotrophic bacterial counts were $<6.5\times10$ for pasteurized milk. All the samples during storage showed the absence of the phosphatase test. This result might be due to proper pasteurization. However, Elmagli and El Zubeir (2006a) demonstrated that 10 % of the pasteurized milk samples were positive to the phosphatase test. No brucella antibodies were detected in pasteurized milk, this might be due to proper pasteurization, and is in accord with that reported by OIE (2005). Moreover this result is better result than that reported by Alves *et al.* (2001). The presence of positive antibodies for brucella in the raw milk samples might suggest infection and/or vaccination, as those herds followed regular vaccination programmes. Similarly negative results of antibiotic residues test were obtained, this may be due to proper follow up of antibiotic withdrawal periods which indicated the good quality of raw milk used. These results are in agreement with Van Schaik *et al.* (2002) and Yamaki *et al.* (2004). It is concluded that the values of chemical contents are within standards limits except for lactose, whose value was lower than the reported by the dairy plant. Low TBC for pasteurized milk was obtained, and the results of this study clearly illustrate that pasteurization plays an important role in the survival and destruction of different bacterial contaminants.

Hazard Analysis Critical Control Point—HACCP

Hazard Analysis Critical Control Point or HACCP is a systematic preventive approach to food safety and pharmaceutical safety that addresses physical, chemical, and biological hazards as a means of prevention rather than finished product inspection. HACCP is used in the food industry to identify potential food safety hazards, so that key actions can be taken to reduce or eliminate the risk of the hazards being realized. The system is used at all stages of food production and preparation processes including packaging, distribution, etc. The Food and Drug Administration (FDA) and the United States Department of Agriculture (USDA) say that their mandatory HACCP programs for juice and meat are an effective approach to food safety and protecting public health. Meat HACCP systems are regulated by the USDA, while seafood and juice are regulated by the FDA. The use of HACCP is currently voluntary in other food industries.

A forerunner to HACCP was developed in the form of production process monitoring during World War II because traditional "end of the pipe" testing was not an efficient way to ferret out artillery shells that would not explode. HACCP itself was conceived in the 1960s when the US National Aeronautics and Space Administration (NASA) asked Pillsbury to design and manufacture the first foods for space flights. Since then, HACCP has been recognized internationally as a logical tool for adapting traditional inspection methods to a modern, science-based, food safety system. Based on risk-assessment, HACCP plans allow both industry and government to allocate their resources efficiently in establishing and auditing safe food production practices. In 1994, the organization of *International HACCP Alliance* was established initially for the US meat and poultry industries to assist them with implementing HACCP and now its membership has been spread over other professional/industrial areas.

Hence, HACCP has been increasingly applied to industries other than food, such as cosmetics and pharmaceuticals. This method, which in effect seeks to plan out unsafe practices, differs from traditional "produce and test" quality control methods which are less successful and inappropriate for highly perishable foods. In the US, HACCP compliance is regulated by 21 CFR part 120 and 123. Similarly, FAO/WHO published a guideline for all governments to handle the issue in small and less developed food businesses.

History

On 4 October 1957, the Soviet Union launched Sputnik, the world's first satellite. American president Dwight D. Eisenhower responded by committing the United States to the space race. Eisenhower signed the National Aeronautics and Space Act on 29 July 1958 that created the National Aeronautics and Space Administration (NASA) to put an American satellite in orbit and to get a person in space.

Food played a critical part in the manned space program. The initial group involved in this were Herbert Hollander, Mary Klicka, and Hamed El-Bisi of the United States Army Laboratories in Natick, Massachusetts and Dr. Paul A. Lachance of the Manned Spaceflight Centre (Johnson Space Centre since February 1973) in Houston, Texas. Pillsbury joined the program as a contractor in 1959 with Howard E. Baumann representing the company as its lead scientist.

The main goal was to produce food that would not crumble under zero gravity, but also be safe to eat. Lachance imposed strict microbial requirements, including pathogen limits (including *E. coli*, *Salmonella*, and *Clostridium botulinum*) on all foods destined for space travel.

All personnel involved realized that traditional quality control methods would be inadequate because there would be so much product testing involved for actual product to be used. NASA own requirements for Critical Control Points (CCP) in engineering management would be used as a guide for food safety. CCP derived from Failure mode and effects analysis (FMEA) from NASA via the munitions industry to test weapon and engineering system reliability.

Using that information, NASA and Pillsbury required contractors to identify "critical failure areas" and eliminate them from the system, a first in the food industry then. Baumann, a microbiologist by training, was so pleased with Pillsbury's experience in the space program that he advocated for his company to adopt what would become HACCP at Pillsbury.

Soon thereafter, Pillsbury was confronted with a food safety issue of its own when glass was found contaminated in farina, a cereal commonly used in infant food. Baumann's leadership promoted HACCP in Pillsbury for producing commercial foods, and applied to its own food production.

This led to a panel discussion at the 1971 National Conference on Food Protection that included examing CCPs and Good Manufacturing Practices in producing safe foods. Several botulism cases were attributed to under-processed low-acid canned foods in 1970-71. The United States Food and Drug Administration (FDA) asked Pillsbury to organize and conduct a training program on the inspection of canned foods for FDA inspectors. This 21 day program was first held in September 1972 with 11 days of classroom lecture and 10 days of canning plant evaluations. Canned food regulations (21 CFR 108, 21 CFR 110, 21 CFR 113, and 21 CFR 114) were first published in 1973. Pillsbury's training program to the FDA in 1972, titled "Food Safety through the Hazard Analysis and Critical Control Point System", was the first time that HACCP was used.

HACCP was initially set on three principles, now shown as principles one, two, and four in the section below. Pillsbury quickly adopted two more principles, numbers three and five, to its own company in 1975. It was further supported by the National Academy of Sciences (NAS) that governmental inspections by the FDA go from reviewing plant records to compliance with its HACCP system. A second proposal by the NAS led to the development of the National Advisory Committee on Microbiological Criteria for Foods (NACMCF) in 1987. NACMCF was initially responsible for defining HACCP's systems and guidelines for its application and were coordinated with the Codex Committee for Food Hygiene, that led to reports starting in 1992 and further harmonization in 1997. By 1997, the seven HACCP principles listed below became the standard. A year earlier, the American Society for Quality offered their first certifications for HACCP Auditors. (First known as Certified Quality Auditor-HACCP, they were changed to Certified HACCP Auditor (CHA) in 2004.

HACCP expanded in all realms of the food industry, going into meat, poultry, seafood, dairy, and has spread now from the farm to the fork.

The HACCP Seven Principles

Principle 1: Conduct a hazard analysis. - Plans determine the food safety hazards and identify the preventive measures the plan can apply to control these hazards. A food safety hazard is any biological, chemical, or physical property that may cause a food to be unsafe for human consumption.

Principle 2: Identify critical control points. - A Critical Control Point (CCP) is a point, step, or procedure in a food manufacturing process at which control can be applied and, as a result, a food safety hazard can be prevented, eliminated, or reduced to an acceptable level.

Principle 3: Establish critical limits for each critical control point. - A critical limit is the maximum or minimum value to which a physical, biological, or chemical hazard must be controlled at a critical control point to prevent, eliminate, or reduce to an acceptable level.

Principle 4: Establish critical control point monitoring requirements. - Monitoring activities are necessary to ensure that the process is under control at each critical control point. In the United States, the FSIS is requiring that each monitoring procedure and its frequency be listed in the HACCP plan.

Principle 5: Establish corrective actions. - These are actions to be taken when monitoring indicates a deviation from an established critical limit. The final rule requires a plant's HACCP plan to identify the corrective actions to be taken if a critical limit is not met. Corrective actions are intended to ensure that no product injurious to health or otherwise adulterated as a result of the deviation enters commerce.

Principle 6: Establish record keeping procedures. - The HACCP regulation requires that all plants maintain certain documents, including its hazard analysis and written HACCP plan, and records documenting the monitoring of critical control points, critical limits, verification activities, and the handling of processing deviations.

Principle 7: Establish procedures for ensuring the HACCP system is working as intended. - Validation ensures that the plants do what they were designed to do; that is, they are successful in ensuring the production of a safe product. Plants will be required to validate their own HACCP plans. FSIS will not approve HACCP plans in advance, but will review them for conformance with the final rule.

Verification ensures the HACCP plan is adequate, that is, working as intended. Verification procedures may include such activities as review of HACCP plans, CCP records, critical limits and microbial sampling and analysis. FSIS is requiring that the HACCP plan include verification tasks to be performed by plant personnel. Verification tasks would also be performed by FSIS inspectors. Both FSIS and industry will undertake microbial testing as one of several verification activities. Verification also includes 'validation' - the process of finding evidence for the accuracy of the HACCP system (e.g. scientific evidence for critical limitations).

Standards

The seven HACCP principles are included in the international standard ISO 22000 FSMS 2005. This standard is a complete food safety and quality management system incorporating the elements of prerequisite programmes (GMP & SSOP), HACCP and the quality management system, which together form an organization's Total Quality Management system.

HACCP Training

HACCP management system trainings are only offered by several commercial enthusiasts. However, ASQ does provide Trained HACCP Auditor (CHA) exam to individuals seeking the professional training. In the UK the Chartered Institute of Environmental Health (CIEH) offer a HACCP for Food Manufacturing qualification accredited by the QCA (Qualifications and Curriculum Authority).

HACCP Application

Applied Range

It can apply to several food categories; sea food, bulk milk production line, Bulk Cream and Butter Production Line, animal meat industry, Organic Chemical Contaminants in Food, Corn Curl Manufacturing Plant and etc.

USA
- Fish and fishery products
- Fresh-cut produces
- Juice and nectary products
- Food outlets
- Meat and poultry products
- School food and services.

HACCP Implementation

It involves monitoring, verifying and validating of the daily work that is compliant with regulatory requirements in all stages all the time. The differences among those three types of work are given by Saskatchewan Agriculture and Food.

HACCP Versus ISO 22000

ISO 22000 is the new standard bound to replace HACCP on issues related to food safety. Although several companies, especially the big ones, have either implemented or are on the point of implementing ISO 22000, there are many others which are rather timid and/or reluctant to implement it. The main reason behind that is the lack of information and the fear that the new standard is too demanding in terms of bureaucratic work, from abstract of case study.

Emulsion

An emulsion is a mixture of two or more immiscible (unblendable) liquids. Emulsions are part of a more general class of two-phase systems of matter called colloids. Although the terms colloid and emulsion are sometimes used interchangeably, emulsion tends to imply that both the dispersed and the continuous phase are liquid. In an emulsion, one liquid (the dispersed phase) is dispersed in the other (the continuous phase).

Examples of emulsions include vinaigrettes, the photo-sensitive side of photographic film, milk and cutting fluid for metal working.

Structure and Properties of Emulsions

It is still common belief that emulsions basically do not display any structure, i.e., the droplets (or in case of dispersions, particles) dispersed in the liquid matrix (the "dispersion medium") are assumed to be statistically distributed. Therefore, for emulsions (like for dispersions) usually percolation theory is assumed to appropriately describe their properties.

However, percolation theory can only be applied if the system it should describe is in or close to thermodynamic equilibrium. There are very few studies about the structure of emulsions (dispersions), although they are plentiful in type and in use all over the world in innumerable applications.

In the following, only such emulsions will be discussed with a dispersed phase diameter of less than 1 μm. To understand the formation and properties of such emulsions (including dispersions), it must be considered, that the dispersed phase exhibits a "surface," which is covered ("wet") by a different "surface" which hence are forming an interface (chemistry). Both surfaces have to be created (which requires a huge amount of energy), and the interfacial tension (difference of surface tension) is not compensating the energy input, if at all.

A review article in introduces into various attempts to describe dispersions / emulsions. Dispersion is a process by which (in the case of solids becoming dispersed in a liquid) agglomerated particles are separated from each other and a new interface, between an inner surface of the liquid dispersion medium and the surface of the particles to be dispersed, is generated. Dispersion is a much more complicated (and less well understood) process than most people believe.

The above cited review article also displays experimental evidence for that dispersions have a structure very much different from any kind of statistical distribution (which would be characteristic for a system in thermodynamic equilibrium, but in contrast very much showing structures similar to self-organisation which can be described by non-equilibrium thermodynamics. This is the reason why some liquid dispersions turn to become gels or even solid at a concentration of a dispersed phase above a certain critical concentration (which is dependant on particle size and interfacial tension).

Also the sudden appearance of conductivity in a system of a dispersed conductive phase in an insulating matrix has been explained. The above cited review article also introduces into some first complete non-equilibrium thermodynamics theory of dispersions.

Appearance and Properties

Emulsions are made up of a dispersed and a continuous phase; the boundary between these phases is called the interface. Emulsions tend to have a cloudy appearance, because the many phase interfaces scatter light that passes through the emulsion.

Emulsions are unstable and thus do not form spontaneously. The basic colour of emulsions is white. If the emulsion is dilute, the Tyndall effect will scatter the light and distort the colour to blue; if it is concentrated, the colour will be distorted towards yellow. This phenomenon is easily observable on comparing skimmed milk (with no or little fat) to cream (high concentration of milk fat). Microemulsions and nanoemulsions tend to appear clear due to the small size of the disperse phase.

Energy input through shaking, stirring, homogenizing, or spray processes are needed to initially form an emulsion. Over time, emulsions tend to revert to the stable state of the phases comprising the emulsion; an example of this is seen in the separation of the oil and vinegar components of Vinaigrette, an unstable emulsion that will quickly separate unless shaken continuously.

Whether an emulsion turns into a water-in-oil emulsion or an oil-in-water emulsion depends on the volume fraction of both phases and on the type of emulsifier. Generally, the Bancroft rule applies: emulsifiers and emulsifying particles tend to promote dispersion of the phase in which they do not dissolve very well; for example, proteins dissolve better in water than in oil and so tend to form oil-in-water emulsions (that is they promote the dispersion of oil droplets throughout a continuous phase of water).

Instability

There are three types of instability: flocculation, creaming, and coalescence. Flocculation describes the process by which the dispersed phase comes out of suspension in flakes. Coalescence is another form of instability, which describes when small droplets combine to form progressively larger ones. Emulsions can also undergo creaming, the migration of one of the substances to the top (or the bottom, depending on the relative densities of the two phases) of the emulsion under the influence of buoyancy or centripetal force when a centrifuge is used.

Surface active substances (surfactants) can increase the kinetic stability of emulsions greatly so that, once formed, the emulsion does not change significantly over years of storage. A Non-Ionic surfactant solution can become self-contained under the force of its own surface tension, remaining in the shape of its previous container for some time after the container is removed. Superfluids flow with zero friction and can escape their containers; an ionic solution tends to retain its current shape.

"Emulsion stability refers to the ability of an emulsion to resist change in its properties over time." D.J. McClements.

Technique Monitoring Physical Stability

Multiple light scattering coupled with vertical scanning is the most widely used technique to monitor the dispersion state of a product, hence identifying and quantifying destabilisation phenomena. It works on concentrated emulsions without dilution. When light is sent through the sample, it is backscattered by the droplets. The backscattering intensity is directly proportional to the size and volume fraction of the dispersed phase. Therefore, local changes in concentration (Creaming) and global changes in size (flocculation, coalescence) are detected and monitored.

Accelerating Methods for Shelf Life Prediction

The kinetic process of destabilisation can be rather long (up to several months or even years for some products) and it is often required for the formulator to use further accelerating methods in order to reach reasonable development time for new product design. Thermal methods are the most commonly used and consists in increasing temperature to accelerate destabilisation (below critical temperatures of phase inversion or chemical degradation). Temperature affects not only the viscosity, but also interfacial tension in the case of non-ionic surfactants or more generally interactions forces inside the system. Storing a dispersion at high temperatures enables to simulate real life conditions for a product (e.g. tube of sunscreen cream in a car in the summer), but also to accelerate destabilisation processes up to 200 times.

Mechanical acceleration including vibration, centrifugation and agitation are sometimes used. They subject the product to different forces that pushes the droplets against one another, hence helping in the film drainage. However, some emulsions would never coalesce in normal gravity, while they do under artificial gravity. Moreover segregation of different populations of particles have been highlighted when using centrifugation and vibration.

Emulsifier

An emulsifier (also known as an emulgent) is a substance which stabilizes an emulsion by increasing its kinetic stability. One class of emulsifiers is known as surface active substances, or surfactants. Examples of food emulsifiers are egg yolk (where the main emulsifying agent is lecithin), honey, and mustard, where a variety of chemicals in the mucilage surrounding the seed hull act as emulsifiers; proteins and low-molecular weight emulsifiers are common as well. Soy lecithin is another emulsifier and thickener. In some cases, particles can stabilize emulsions as well through a mechanism called Pickering stabilization. Both mayonnaise and Hollandaise sauce are oil-in-water emulsions that are stabilized with egg yolk lecithin or other types of food additives such as Sodium stearoyl lactylate.

Detergents are another class of surfactant, and will physically interact with both oil and water, thus stabilizing the interface between oil or water droplets in suspension. This principle is exploited in soap to remove grease for the purpose of cleaning. A wide variety of emulsifiers are used in pharmacy to prepare emulsions such as creams and lotions. Common examples include emulsifying wax, cetearyl alcohol, polysorbate 20, and ceteareth 20. Sometimes the inner phase itself can act as an emulsifier, and the result is nanoemulsion - the inner state disperses into nano-size droplets within the outer phase. A well-known example of this phenomenon, the ouzo effect, happens when water is poured in a strong alcoholic anise-based beverage, such as ouzo, pastis, arak or raki. The anisolic compounds, which are soluble in ethanol, now form nano-sized droplets and emulgate within the water. The colour of such diluted drink is opaque and milky.

In Food

Oil-in-water emulsions are common in food. Notable examples include:

- Crema in espresso – coffee oil in water (brewed coffee), unstable
- Hollandaise sauce – similar to mayonnaise
- Mayonnaise – vegetable oil in lemon juice or vinegar, with egg yolk lecithin as emulsifier
- Vinaigrette – vegetable oil in vinegar; if prepared with only oil and vinegar (without an emulsifier), yields an unstable emulsion.

In Medicine

In pharmaceutics, hairstyling, personal hygiene and cosmetics, emulsions are frequently used. These are usually oil and water emulsions, but which is dispersed and which is continuous depends on the pharmaceutical formulation. These emulsions may be called creams, ointments, liniments (balms), pastes, films or liquids, depending mostly on their oil and water proportions and their route of administration. The first 5 are topical dosage forms, and may be used on the surface of the skin, transdermally, ophthalmically, rectally or vaginally. A very liquidy emulsion may also be used orally, or it may be injected using various routes (typically intravenously or intramuscularly). Popular medicated emulsions include calamine lotion, cod liver oil, Polysporin, cortisol cream, Canesten and Fleet.

Microemulsions are used to deliver vaccines and kill microbes. Typically, the emulsions used in these techniques are nanoemulsions of soybean oil, with particles that are 400-600 nm in diameter. The process is not chemical, as with other types of antimicrobial treatments, but mechanical. The smaller the droplet, the greater the surface tension and thus the greater the force to merge with other lipids.

The oil is emulsified using a high shear mixer with detergents to stabilize the emulsion, so when they encounter the lipids in the membrane or envelope of bacteria or viruses, they force the lipids to merge with themselves. On a mass scale, this effectively disintegrates the membrane and kills the pathogen.

This soybean oil emulsion does not harm normal human cells nor the cells of most other higher organisms. The exceptions are sperm cells and blood cells, which are vulnerable to nanoemulsions due to their membrane structures. For this reason, these nanoemulsions are not currently used intravenously. The most effective application of this type of nanoemulsion is for the disinfection of surfaces. Some types of nanoemulsions have been shown to effectively destroy HIV-1 and various tuberculosis pathogens, for example, on non-porous surfaces.

In Fire Fighting

Emulsifying agents are effective at extinguishing fires on small thin layer spills of flammable liquids (Class B fires). Extinguishment is achieved by encapsulating the fuel in a fuel-water emulsion thereby trapping the flammable vapors in the water phase. This emulsion is achieved by applying an aqueous surfactant solution to the fuel through a high pressure nozzle.

Emulsifiers are not effective at extinguishing large Class B fuel in depth fires. This is because the amount of agent needed for extinguishment is a function of the volume of the fuel whereas agents such as aqueous film forming foam (AFFF) need only cover the surface of the fuel to achieve vapor mitigation.

Uses

Emulsions are mainly used in many major chemical industries. In the pharmaceutical industry they are used to make medicines with a more appealing flavour and to improve value by controlling the amount of active ingredients. The most widely-used emulsions are non-ionic because they have low toxicity, but cationic emulsions are also used in some products because of their antimicrobial properties. Emulsions are also used in making many hair and skin products, such as various types of oils and waxes.

Shelf-life Predicting Methods for Milk

New analysis allows prediction of shelf life for pasteurized bottled milk. At I&A Lab we think that the most important part of a plant is production, and the laboratory should provide help to production to solve and prevent the development of quality problems. Production faces different types of problems like:

1- Quality of the raw milk received
2- Equipment
3- CIP
4- Packing material
5- Personnel.

All of these can affect the quality of the final product and as a consequence the shelf life.

Our analysis is designed for use with pasteurized milk. To establish the shelf life of bottled milk it is necessary to be able to predict how the product is going to behave in the stores, at the correct temperature, and to know what is going to happen if the product is abused. The ideal is to have this prediction in a reasonable amount of time, the sooner after the milk is bottled being better.

There are different groups of microorganisms capable of growing and spoiling the milk, for our study we consider that the mesophilic (growing between 68° and 113°F), and psychrophilic (growing between 20° and 68°F) are the most important groups, so we focus our effort on the mesophilic's which can spoil the milk when it is abused, and the psychrophilic's which can spoil the milk even at temperatures below 45°F.

Most psychrophilic microorganisms are a result of post - pasteurization contamination, due to the fact that they usually die with pasteurization. It is known that milk is an excellent media for the growth of many microorganisms, it is also known that if the milk is kept under 45°F many microorganisms stop or decrease their growth.

The legal analysis for pasteurized bottled milk uses one milliliter of milk in a solid media for total aerobic counts and coliform counts. It is not a requirement to analyze psychrophilics, but the most widely used analysis takes 7 to 10 days to complete.

The solid media employed is very different from liquid milk, so much so that it takes 48 hrs to see colonies growing in the solid media plates, incubated at 113°F, while it takes only a few hours for the milk to spoil. This means that for most of the microorganisms it is easier to grow in a liquid than in a solid. The reason for the use of solid media is to be able to count the amount of colonies per milliliter of milk.

We have developed a new method using liquid milk in a volume 10 times higher than normal in special equipment using new software. This allows us to detect microorganisms in very low concentrations. We also employ two different medias for psychrophilic bacteria, along with media for pseudomonas, and mesophilic bacteria. This allows us to have a wide spectrum of detectable milk inhabiting microorganisms.

We created a database with our results and compared them against the results of the plated fresh samples, and the same samples preincubated at 68°F for 16 hrs. Finally, the milk was flavoured over a period of time until it was determined to be no longer acceptable for consumption.

After more than 8000 samples being analyzed we are able to detect in 20 hrs samples which will have 10 days or less of shelf life. If the milk is contaminated for any reason we will be able to give the production plant this information. We are able to report in 24 hrs if the milk has any problem which may compromise its shelf life.

We will issue a second report in 48 hrs stating the amount of psychrophilic, pseudomonas, coliform, and mesophilic bacteria in the milk sample, with a prediction, in number of days, with which the milk will be in good condition if stored at 45°F. Also, the number of days of shelf life if the milk is kept between 50° and 55°F will be supplied.

During our study we found a type of bacteria capable of growing at 68°F. We identified this as Bacillus megaterium (mesophilic). As its presence occurred frequently we studied it and determined that even though it is capable of growing at room temperatures it is not capable of growing below 45°F, however it grows rapidly if the milk is abused.

As the milk is a biological product, and there are millions of different microorganisms capable of growing and making changes in the milk, we will continue our study. In the future we will be able to provide more information to help production managers make better decisions regarding the shelf life of their products.

Determination of the End of Shelf Life for Milk
Using Weibull Hazard Method

Undesirable changes in dairy products may be instigated by microbial growth and metabolism or by chemical reactions. The determinants of shelf life of fresh dairy products are usually the spoilage bacteria that have the ability to grow at refrigerated temperatures.

This microbial growth induces changes in the taste and odour of milk such as sour, putrid, bitter, malty, fruity, rancid and unclean. In addition, psychrotrophs which are common contaminants in milk, synthesize enzymes, many of which survive the pasteurization heat treatment and during storage biochemically alter the milk and eventually cause spoilage.

Growth of psychrotrophic bacteria is predominantly responsible for influencing the keeping quality of milk and dairy products held below 7°C. Raw and pasteurized milk usually spoils when held at refrigeration temperatures because of the effects of psychrotrophic contaminates.

The populations of microorganisms needed to cause detectable changes in milk varies among genera and species within a genus, but levels at which flavour changes occur are similar at 6 and 20°C. Milk spoilage by psychrotrophs was reported in the range of populations of 1×10^2 to 1×10^9 per ml. It is therefore unclear whether pschrotrophs counts can be used as an index in the determination of milk quality or shelf life from a sensory standpoint.

As noted earlier, microbial spoilage leads to sensory deterioration of the milk. It may therefore be suggested that the microbial quality of the milk should correlate well to its sensory end of shelf life. The end of shelf life can be determined from sensory data by various graphical methods. The use of hazard rate for shelf life testing of food was introduced by Gacula (1975). Using this method, one can determine the end of shelf life according to the percent of customers a company is prepared to displease.

The maximum likelihood graphical procedure, or Weibull Hazard method has been used for shelf life of luncheon meats, oat bran cereal, ice cream, cottage cheese, Bockwurst sausages and butter, and other food products.

The objective of this research was to determine whether or not a consumer determined end of sensory shelf life can be described by some microbial index regardless of the temperature conditions that the milk is stored at.

Materials and Methods

Milk

The milk used in this study was TLC‴ fat free milk with added Calcium. This milk is also fortified with nonfat milk proteins. The raw milk was held for no longer than 48 hours at 2°C before processing and then pasteurized (20 s, 80°C). Half gallon, paper board cartons of milk were picked up within 2 h after bottling, taken off the production line consecutively by a plant supervisor, in order to minimize variability between cartons. The cartons of milk were transported to the University of Minnesota on ice in a cooler. Immediately after arrival, the milk was sampled for microbial quality and tasted by three expert dairy panelists to ensure that the milk was of good quality.

Microbial Counts

Total aerobic bacteria as well as psychrotrophic bacteria were enumerated using 3M Petrifilm (3M Co., St. Paul, MN). Samples were diluted in 0.1% peptone, and 1 ml of sample was transfered onto the film in duplicate. The Petrifilm contained standard method nutrients and a cold water soluble gelling agent (8, 20). The bottom film is coated with nutrients and gelling agent, while the top film is coated with the gelling agent and 2,3, 5-triphenylterazoluim chloride (TTC). Colonies appear red and were counted following incubation at 37°C for 48 h for total aerobic or 10 d at 7°C for pyschrotrophs.

Microbial Growth

The growth of total aerobic bacteria and psychrotrophic bacteria in the milk was measured at five constant temperatures: 2, 5, 7, 12, and 15°C (± 1°C). A TempTale (Sensitech, Beverly, MA) temperature recorder, placed in the coolers along with the milk verified the temperature history. Samples were drawn at predetermined intervals according to the storage temperature conditions. The lag times were determined graphically, and the growth rate constant was calculated through linear regression of the exponential phase of the growth curves.

Sensory Testing

The sensory testing was carried out following the Weibull Hazard method, where the initial number of panelists was n0 = 3 and the constant with which the number of panelists was increased for each subsequent test was nc = 1. The interval between sensory testing was predetermined for each of the five different storage conditions.

Because the spoilage of milk at 14°C occurred at an accelerated rate, sensory samples were held overnight so that the sensory test could be carried out at a convenient time for the panelists. The panelists were prescreened and were required to meet the criterion that they consume at least one 8-ounce glass of milk a day. A pool of 33 panelists who met this requirement, 16 male and 17 female, ranging in age form 18 to 45 were available for sensory testing. The panelists were financially compensated according to the number of samples that they tested.

For each sensory test a sample of milk was taken from the milk cartons, poured into a glass flask and the flask was immediately placed into an ice bath to slow down any microbial growth that might have caused any further deterioration of the sensory quality of the milk. Approximately 10 ml of milk was poured into cups that were labelled with random three digit numbers for identification purposes. A tray of milk samples for each of the panelists was prepared approximately half an hour to an hour before sensory testing took place.

To ensure that the samples were all the same temperature when the panelists received their trays the trays were stored in conventional home refrigerators held at 4°C. The trays, consisted of samples of milk from the different storage temperatures. The trays were presented to the panelists in sensory booths where the sensory testing was held. Panelists were asked to taste the first sample and determine whether the milk was acceptable or unacceptable, where a response of acceptable implied that the panelist would be willing to drink an entire glass of the sample. Panelists were asked to wait two minutes between samples and to rinse their mouths with water in between.

The end of sensory shelf life was determined at 69.3% cumulative hazard or a critical probability of 50%. The data was regressed using the least squares method up to 100% cumulative hazard.

Results and Discussion

Microbial Growth

The growth of the total aerobic bacteria and the psychrotrophic bacteria were obtained at 2, 5, 7, 12, and 14°C (± 1°C).

The total aerobic microbial population exhibited typical growth curves at all temperatures, however, at 2 and 5°C the milk reached the end of sensory shelf life before the end of the lag phase. The psychrotrophic bacteria, exhibit distinct lag and log growth phases at 5, 7, 12, and 14°C. At 2°C however, a rapid growth of psychrotrophs occurred immediately following the opening of the milk carton for sampling, thus, obtaining a growth curve for psychrotrophs at 2°C was not practical within this experimental setup. It is likely that the rapid growth following opening of the milk carton was due to the change in available oxygen. Once the carton is opened the amount of available oxygen for the microorganisms in the milk increases and facilitates their growth. Sinclair and Stokes (1963) support this explanation with the discovery that in general, due to an increase in the availability of oxygen higher counts are observed.

The growth of both aerobic and pschrotrophs populations at 5°C did not exhibit a distinct logarithmic phase. The absence of a logarithmic growth phase at 5°C can be attributed to the fact that the samples were taken from more than one carton throughout the experiment. Thus the variability in the population of microorganisms from carton to carton may result in difficulties in detecting a distinct lag and exponential phase.

The cartons were taken from the production line in consecutive order so that the cartons could be considered to be identical and so the sampling from the cartons during the growth curve study could be made randomly between the opened cartons. However, Maxcy and Wallen (1983) found that heterogeneity between cartons was apparent even when samples were taken sequentially from a single production line.

Growth Parameters

It was not possible to determine the duration of the lag phase for psychrotrophic bacteria of the milk stored at 2°C because the psychrotrophic counts showed no distinctive pattern. Since the milk stored at 5° and 2°C reached the sensory end point during the lag phase of both the total aerobic bacteria and psychrotrophic bacteria, the exponential growth rates for the bacteria at these two storage temperatures, were not calculated. Data presented by Fu (1989) showed that the temperature dependence of the growth rate constants and the lag time for microbial growth in a model milk system fits the Arrhenius model:

Indigenous (Indian) Dairy Products

A variety of dairy projects are indigenous to India and an important part of Indian cuisine. The majority of these products can be broadly classified into curdled products, like chhena, or non-curdled products, like khoa.

Curdled Dairy Products

- Paneer is an unaged, acid-set, non-melting farmer cheese made by curdling heated milk with lemon juice or other non-rennet food acid, and then removing the whey and pressing the result into a dry unit.
- Chhena is like paneer, except some whey is left and the mixture is beaten thoroughly until it becomes soft, of smooth consistency, and malleable but firm.
- Sandesh is a confection made from chhena mixed with sugar then grilled lightly to caramelize, but removed from heat and molded into a ball or some shape.
- Rasgulla is a confection made from mixture of chhena and semolina rolled into a ball and boiled in syrup.

Non-curdled Dairy Products

- Khoa or Mawa is made by reducing milk in an open pan over heat.
- Peda is a confection made by mixing sugar with khoa and adding flavoring, such as cardamom.
- Barfi is a confection made by reducing milk and sugar until it solidifies and adding flavoring, such as pistachio.
- Gulab jamun is a confection made by mixing khoa and sugar, caramelizing it by frying, and soaking it in syrup containing rosewater.
- Kulfi is made from slowly freezing sweetened condensed milk. In comparison to ice cream, kulfi is not whipped or otherwise aerated.

- Ghee is type of clarified butter that is cooked long enough to caramelize the milk sugar and sterilize the liquid.

Fermented Dairy Products
- Mishti doi is *dahi* (Indian yogurt) mixed with sugar
- Shrikhand is strained yoghurt mixed with sugar, and often flavourings such as cardamom, saffron, or fruit.
- Wheyvit is an alcoholic beverage prepared by fermenting whey with yeast.

Other Dairy Products
- Kheer is made by boiling rice or broken wheat with milk and sugar, and sometimes flavoured with cardamom, raisins, saffron, pistachios, or almonds.
- Chhena Murki is made by frying cubes of chhena to burn the outside, then soaking them in syrup flavoured with cardamom.
- Pantooa is like gulab jamun, except with some chhena mixed with the usual ingredients.

4
Chemistry of Dairy Products

Physical Status of Milk

About 87% of milk is water, in which the other constituents are distributed in various forms. We distinguish among several kinds of distribution according to the type and size of particle present in the liquid.

Kind of solution	Particle diameter (nm)
Ionic solution	0.01–1
Molecular solution	0.1–1
Colloid (fine dispersion)	1–100
Coarse dispersion (suspension or emulsion)	50–100

In milk we find examples of emulsions, colloids, molecular and ionic solutions.

Ionic Solutions

An ionic solution is obtained when the forces that hold the ions together in a solid salt are overcome. The dissolved salt breaks up into ions which float freely in the solvent. Thus when common salt—sodium chloride—is dissolved in water it becomes an ionic solution of free sodium and chloride ions. Ionic solutions are largely of inorganic compounds.

Molecular Solutions

In a molecular solution the molecules are only partly, if at all, dissociated into ions. The degree of dissociation represents an equilibrium which is influenced by other substances in the solution and by the pH (or hydrogen ion concentration) of the solution. Molecular solutions are usually of organic compounds.

Colloids

In a colloid, one substance is dispersed in another in a finer state than an emulsion but the particle size is larger than that in a true solution. Colloidal systems are classified according to the physical state of the two phases. In a colloid, solid particles consisting of groups of molecules float freely. The particles in a colloid are much smaller than those in a suspension and a colloid is much more stable.

Emulsions

An emulsion consists of one immiscible liquid dispersed in another in the form of droplets—the disperse phase. The other phase is referred to as the continuous phase. The systems have minimal stability and require the presence of a surface-active or emulsifying agent for stability. In foods, emulsions usually contain oil and water. If water is the continuous phase and oil the disperse phase, it is an oil-in-water (o/w) emulsion, e.g. milk or cream. In the reverse case the emulsion is a water-in-oil (w/o) type, e.g. butter. In summary, an emulsion consists of three elements, the continuous phase, the disperse phase and the emulsifying agent.

Dispersions

A dispersion is obtained when particles of a substance are dispersed in a liquid. A suspension consists of solid particles dispersed in a liquid, and the force of gravity can cause them to sink to the bottom or float to the top. For example, fine sand, dispersed in water, soon settles out.

PH and Acidity

An acid is a substance which dissociates to produce hydrogen ions in solution. A base (alkaline) is a substance which produces hydroxyl ions in solution. It can equally be stated that an acid is a substance which donates a proton and a base is a substance which accepts a proton. The symbol pH is used to denote acidity; it is inversely related to hydrogen ion concentration.

Neutrality is pH 7

Acidity is less than pH 7

Alkalinity is more than pH 7

Fresh milk has a pH of 6.7 and is therefore slightly acidic.

When an acid is mixed with a base, neutralisation takes place; similarly a base will be neutralised by an acid.

Buffer Solutions

Buffers are defined as materials that resist a change in pH on addition of acid or alkali. Characteristically they consist of a weak acid or a weak base and its salt. Milk contains a large number of these substances and consequently behaves as a buffer solution. Fresh cows milk has a pH of between 6.7 and 6.5. Values higher than 6.7 denote mastitic milk and values below pH 6.5 denote the presence of colostrum or bacterial deterioration. Because milk is a buffer solution, considerable acid development may occur before the pH changes. A pH lower than 6.5 therefore indicates that considerable acid development has taken place. This is normally due to bacterial activity. Litmus test papers, which indicate pH, are used to test milk activity; pH measurements are often used as acceptance tests for milk.

Measuring milk acidity is an important test used to determine milk quality. Acidity measurements are also used to monitor processes such as cheese-making and yoghurt-making. The titratable acidity of fresh milk is expressed in terms of percentage lactic acid, because lactic acid is the principal acid produced by fermentation after milk is drawn from the udder and fresh milk contains only traces of lactic acid. However, due to the buffering capacity of the proteins and milk salts, fresh milk normally exhibits an initial acidity of 0.14 to 0.16% when titrated using sodium hydroxide to a phenolphthalein end-point.

Milk Constituents

The quantities of the main milk constituents can vary considerably depending on the individual animal, its breed, stage of lactation, age and health status. Herd management practices and environmental conditions also influence milk composition. The average composition of cows milk is shown in Table 1.

Table 1: Composition of cows milk

Main constituent	Range(%)	Mean(%)
Water	85.5 – 89.5	87.0
Total solids	10.5 – 14.5	13.0
Fat	2.5 – 6.0	4.0
Proteins	2.9 – 5.0	3.4
Lactose	3.6 – 5.5	4.8
Minerals	0.6 – 0.9	0.8

Water is the main constituent of milk and much milk processing is designed to remove water from milk or reduce the moisture content of the product.

Milk Fat

If milk is left to stand, a layer of cream forms on the surface. The cream differs considerably in appearance from the lower layer of skim milk. Under the microscope cream can be seen to consist of a large number of spheres of varying sizes floating in the milk.

Each sphere is surrounded by a thin skin—the fat globule membrane—which acts as the emulsifying agent for the fat suspended in milk. The membrane protects the fat from enzymes and prevents the globules coalescing into butter grains. The fat is present as an oil-in-water emulsion: this emulsion can be broken by mechanical action such as shaking.

Fats are partly solid at room temperature. The term oil is reserved for fats that are completely liquid at room temperature. Fats and oils are soluble in non-polar solvents, e.g. ether. About 98% of milk fat is a mixture of triacyl glycerides.

There are also neutral lipids, fat-soluble vitamins and pigments (e.g. carotene, which gives butter its yellow colour), sterols and waxes. Fats supply the body with a concentrated source of energy: oxidation of fat in the body yields 9 calories/g. Milk fat acts as a solvent for the fat-soluble vitamins A, D, E and K and also supplies essential fatty acids (linoleic, linolenic and arachidonic).

A fatty-acid molecule comprises a hydrocarbon chain and a carboxyl group (-COOH). In saturated fatty acids the carbon atoms are linked in a chain by single bonds.

In unsaturated fatty acids there is one double bond and in poly-unsaturated fatty acids there is more than one double bond. Fatty acids vary in chain length from 4 carbon atoms, as in butyric acid (found only in butterfat), to 20 carbon atoms, as in arachidonic acid. Nearly all the fatty acids in milk contain an even number of carbon atoms.

Fatty acids can also vary in degree of unsaturation, e.g. C18:0 stearic (saturated), C18:1 oleic (one double bond), C18:2 linoleic (two double bonds), C18:3 linolenic (three double bonds).

Table 2: Principal fatty acids found in milk triglycerides.

	Molecular formula	Chain length	Melting point
Butyric	$CH_3(CH_2)_2COOH$	C_4	$-8°C$
Caproic	$CH_3(CH_2)_4COOH$	C_6	$-2°C$
Caprylic	$CH_2(CH_2)_6COOH$	C_8	$16°C$
Capric	$CH_3(CH_2)_8COOH$	C_{10}	$31.5°C$
Lauric	$CH_3(CH_2)_{10}COOH$	C_{12}	$44°C$
Myristic	$CH_3(CH_2)_{12}COOH$	C_{14}	$58°C$
Palmitic	$CH_3(CH_2)_{14}COOH$	C_{16}	$64°C$
Stearic	$CH_3(CH_2)_{16}COOH$	C_{18}	$70°C$
Arichidonic	$CH_3(CH_2)_{18}COOH$	C_{20}	
Oleic	$CH_3(CH_2)_{7CH}=CH(CH_2)_7COOH$	$C_{18:1}$	$13°C$
Linoleic	$CH_3(CH_2)_4(CH=CH.CH_2)_2(CH_2)_6COOH$	$C_{18:2}$	$-5°C$
Linolenic	$CH_3.CH_2(CH=CH.CH_2)_3(CH_2)_6COOH$	$C_{18:3}$	

The melting point and hardness of the fatty acid is affected by:
- the length of the carbon chain, and
- the degree of unsaturation.

As chain length increases, melting point increases. As the degree of unsaturation increases, the melting point decreases. Fats composed of short-chain, unsaturated fatty acids have low melting points and are liquid at room temperature, i.e. oils. Fats high in long-chain saturated fatty acids have high melting points and are solid at room temperature. Butterfat is a mixture of fatty acids with different melting points, and therefore does not have a distinct melting point. Since butterfat melts gradually over the temperature range of 0–40°C, some of the fat is liquid and some solid at temperatures between 16 and 25°C. The ratio of solid to liquid fat at the time of churning influences the rate of churning and the yield and quality of butter.

Fats readily absorb flavours. For example, butter made in a smoked gourd has a smokey flavour. Fats in foods are subject to two types of deterioration that affect the flavour of food products.

1. *Hydrolytic rancidity:* In hydrolytic rancidity, fatty acids are broken off from the glycerol molecule by lipase enzymes produced by milk bacteria. The resulting free fatty acids are volatile and contribute significantly to the flavour of the product.
2. *Oxidative rancidity:* Oxidative rancidity occurs when fatty acids are oxidised. In milk products it causes tallowy flavours. Oxidative rancidity of dry butterfat causes off-flavours in recombined milk.

Milk Proteins

Proteins are an extremely important class of naturally occurring compounds that are essential to all life processes. They perform a variety of functions in living organisms ranging from providing structure to reproduction. Milk proteins represent one of the greatest contributions of milk to human nutrition. Proteins are polymers of amino acids. Only 20 different amino acids occur, regularly in proteins. They have the general structure:

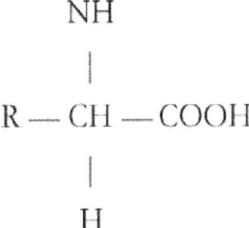

R represents the organic radical. Each amino acid has a different radical and this affects the properties of the acid. The content and sequence of amino acids in a protein therefore affect its properties. Some proteins contain substances other than amino acids, e.g. lipoproteins contain fat and protein. Such proteins are called conjugated proteins:

Phosphoproteins: Phosphate is linked chemically to these proteins—examples include casein in milk and phosphoproteins in egg yolk.

Lipoproteins: These combinations of lipid and protein are excellent emulsifying agents. Lipoproteins are found in milk and egg yolk.

Chromoproteins: These are proteins with a coloured prosthetic group and include haemoglobin and myoglobin.

Whey Proteins

After the fat and casein have been removed from milk, one is left with whey, which contains the soluble milk salts, milk sugar and the remainder of the milk proteins. Like the proteins in eggs, whey proteins can be coagulated by heat. When coagulated, they can be recovered with caseins in the manufacture of acid-type cheeses. The whey proteins are made up of a number of distinct proteins, the most important of which are β-lactoglobulin and lactoglobulin. β-lactoglobulin accounts for about 50% of the whey proteins, and has a high content of essential amino acids. It forms a complex with K-casein when milk is heated to more than 75°C, and this complex affects the functional properties of milk. Denaturation of β-lactoglobulin causes the cooked flavour of heated milk.

Casein

Casein was first separated from milk in 1830, by adding acid to milk, thus establishing its existence as a distinct protein. In 1895 the whey proteins were separated into globulin and albumin fractions.

It was subsequently shown that casein is made up of a number of fractions and is therefore heterogeneous. The whey proteins are also made up of a number of distinct proteins as shown in the scheme in Figure.

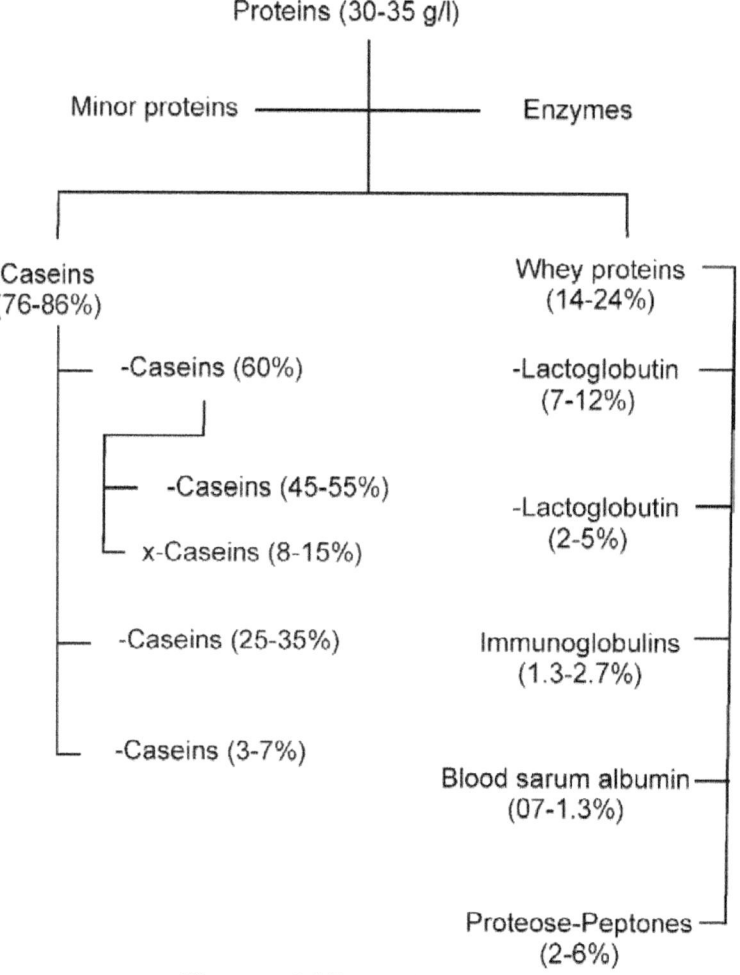

Figure 1: Milk protein fractions

Casein is easily separated from milk, either by acid precipitation or by adding rennin. In cheese-making most of the casein is recovered with the milk fat. Casein can also be recovered from skim milk as a separate product. Casein is dispersed in milk in the form of micelles. The micelles are stabilised by the K-casein. Caseins are hydrophobic but K-casein contains a hydrophilic portion known as the glycomacropeptide and it is this that stabilises the micelles. The structure of the micelles is not fully understood.

When the pH of milk is changed, the acidic or basic groups of the proteins will be neutralised. At the pH at which the positive charge on a protein equals exactly the negative charge, the net total charge of the protein is zero. This pH is called the isoelectric point of the protein (pH 4.6 for casein). If an acid is added to milk, or if acid-producing bacteria are allowed to grow in milk, the pH falls. As the pH falls the charge on casein falls and it precipitates. Hence milk curdles as it sours, or the casein precipitates more completely at low pH.

Other Milk Proteins

In addition to the major protein fractions outlined, milk contains a number of enzymes. The main enzymes present are lipases, which cause rancidity, particularly in homogenised milk, and phosphatase enzymes, which catalyse the hydrolysis of organic phosphates. Measuring the inactivation of alkaline phosphatase is a method of testing the effectiveness of pasteurisation of milk.

Peroxidase enzymes, which catalyse the breakdown of hydrogen peroxide to water and oxygen, are also present. Lactoperoxidase can be activated and use is made of this for milk preservation. Milk also contains protease enzymes, which catalyse the hydrolysis of proteins, and lactalbumin, bovine serum albumin, the immune globulins and lactoferrin, which protect the young calf against infection.

Milk Carbohydrates

Lactose is the major carbohydrate fraction in milk. It is made up of two sugars, glucose and galactose (Figure). The average lactose content of milk varies between 4.7 and 4.9%, though milk from individual cows may vary more. Mastitis reduces lactose secretion.

Lactose is a source of energy for the young calf, and provides 4 calories/g of lactose metabolised. It is less soluble in water than sucrose and is also less sweet. It can be broken down to glucose and galactose by bacteria that have the enzyme β-galactosidase. The glucose and galactose can then be fermented to lactic acid. This occurs when milk goes sour. Under controlled conditions they can also be fermented to other acids to give a desired flavour, such as propionic acid fermentation in Swiss-cheese manufacture.

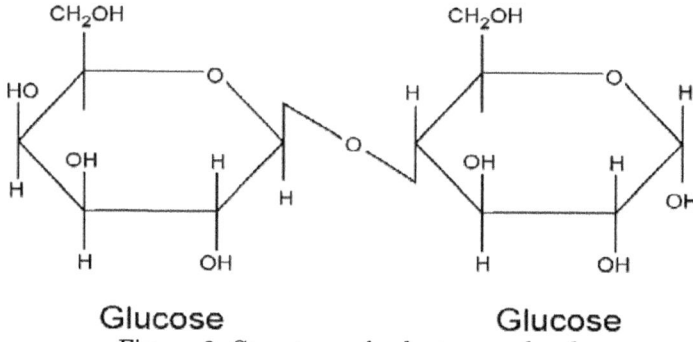

Figure 2: Structure of a lactose molecule

Lactose is present in milk in molecular solution. In cheese-making lactose remains in the whey fraction. It has been recovered from whey for use in the pharmaceutical industry, where its low solubility in water makes it suitable for coating tablets.

It is used to fortify baby-food formula. Lactose can be sprayed on silage to increase the rate of acid development in silage fermentation. It can be converted into ethanol using certain strains of yeast, and the yeast biomass recovered and used as animal feed. However, these processes are expensive and a large throughput is necessary for them to be profitable. For smallholders, whey is best used as a food without any further processing.

Heating milk to above 100°C causes lactose to combine irreversibly with the milk proteins. This reduces the nutritional value of the milk and also turns it brown.

Because lactose is not as soluble in water as sucrose, adding sucrose to milk forces lactose out of solution and it crystallises. This causes sandiness in such products as ice cream. Special processing is required to crystallise lactose when manufacturing products such as instant skim milk powders. Some people are unable to metabolise lactose and suffer from an allergy as a result. Pre-treatment of milk with lactase enzyme breaks down the lactose and helps overcome this difficulty. In addition to lactose, milk contains traces of glucose and galactose. Carbohydrates are also present in association with protein.

K-casein, which stabilises the casein system, is a carbohydrate-containing protein.

Minor Milk Constituents

In addition to the major constituents discussed above, milk also contains a number of organic and inorganic compounds in small or trace amounts, some of which affect both the processing and nutritional properties of milk.

Milk Salts

Milk salts are mainly chlorides, phosphates and citrates of sodium, calcium and magnesium. Although salts comprise less than 1 % of the milk they influence its rate of coagulation and other functional properties. Some salts are present in true solution. The physical state of other salts is not fully understood. Calcium, magnesium, phosphorous and citrate are distributed between the soluble and colloidal phases. Their equilibria are altered by heating, cooling and by a change in pH.

Table 3: Distribution of milk salts between the soluble and colloidal phases.

	Total (mg/100 ml of milk)	Dissolved	Colloidal
Calcium	1320.1	51.8	80.3
Magnesium	10.8	7.9	2.9
Total phosphorus	95.8	36.3	59.6
Citrate	156.6	141.6	15.0

In addition to the major salts, milk also contains trace elements. Some elements come to the milk from feeds, but milking utensils and equipment are important sources of such elements as copper, iron, nickel and zinc.

Milk Vitamins

Milk contains the fat-soluble vitamins A, D, E and K in association with the fat fraction and water-soluble vitamins B complex and C in association with the water phase. Vitamins are unstable and processing can therefore reduce the effective vitamin content of milk.

Milk Protein

General Protein Definition and Chemistry

Proteins are chains of amino acid molecules connected by peptide bonds.

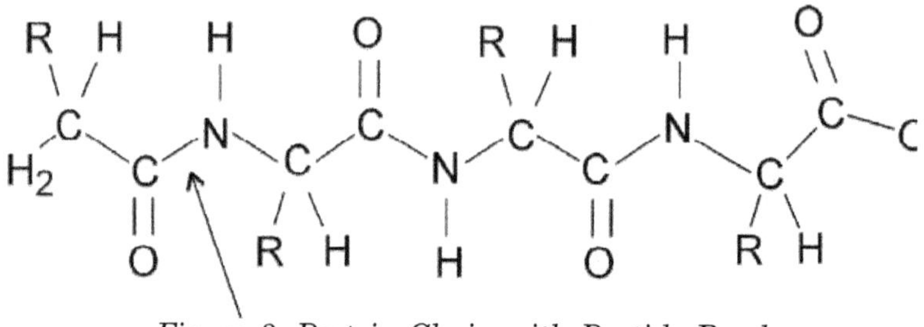

Figure 3: Protein Chain with Peptide Bond
R= amino acid group

There are many types of proteins and each has its own amino acid sequence (typically containing hundreds of amino acids). There are 22 different amino acids that can be combined to form protein chains. There are 9 amino acids that the human body cannot make and must be obtained from the diet. These are called the essential amino acids.

The amino acids within protein chains can bond across the chain and fold to form 3-dimensional structures. Proteins can be relatively straight or form tightly compacted globules or be somewhere in between. The term "denatured" is used when proteins unfold from their native chain or globular shape. Denaturing proteins is beneficial in some instances, such as allowing easy access to the protein chain by enzymes for digestion, or for increasing the ability of the whey proteins to bind water and provide a desirable texture in yogurt production.

Milk Protein Chemistry

Milk contains 3.3% total protein. Milk proteins contain all 9 essential amino acids required by humans. Milk proteins are synthesized in the mammary gland, but 60% of the amino acids used to build the proteins are obtained from the cow's diet. Total milk protein content and amino acid composition varies with cow breed and individual animal genetics.

There are 2 major categories of milk protein that are broadly defined by their chemical composition and physical properties. The casein family contains phosphorus and will coagulate or precipitate at pH 4.6. The serum (whey) proteins do not contain phosphorus, and these proteins remain in solution in milk at pH 4.6. The principle of coagulation, or curd formation, at reduced pH is the basis for cheese curd formation. In cow's milk, approximately 82% of milk protein is casein and the remaining 18% is serum, or whey protein.

The casein family of protein consists of several types of caseins (α-s1, α-s2, β, and 6) and each has its own amino acid composition, genetic variations, and functional properties. The caseins are suspended in milk in a complex called a micelle that is discussed below in the physical properties section. The caseins have a relatively random, open structure due to the amino acid composition (high proline content).

The high phosphate content of the casein family allows it to associate with calcium and form calcium phosphate salts. The abundance of phosphate allows milk to contain much more calcium than would be possible if all the calcium were dissolved in solution, thus casein proteins provide a good source of calcium for milk consumers. The 6-casein is made of a carbohydrate portion attached to the protein chain and is located near the outside surface of the casein micelle. In cheese manufacture, the 6-casein is cleaved between certain amino acids, and this results in a protein fragment that does not contain the amino acid phenylalanine. This fragment is called milk glycomacropeptide and is a unique source of protein for people with phenylketonuria.

The serum (whey) protein family consists of approximately 50% β-lactoglobulin, 20% α-lactalbumin, blood serum albumin, immunoglobulins, lactoferrin, transferrin, and many minor proteins and enzymes. Like the other major milk components, each whey protein has its own characteristic composition and variations. Whey proteins do not contain phosphorus, by definition, but do contain a large amount of sulfur-containing amino acids.

These form disulfide bonds within the protein causing the chain to form a compact spherical shape. The disulfide bonds can be broken, leading to loss of compact structure, a process called denaturing. Denaturation is an advantage in yogurt production because it increases the amount of water that the proteins can bind, which improves the texture of yogurt. This principle is also used to create specialized whey protein ingredients with unique functional properties for use in foods. One example is the use of whey proteins to bind water in meat and sausage products.

The functions of many whey proteins are not clearly defined, and they may not have a specific function in milk but may be an artifact of milk synthesis. The function of β-lactoglobulin is thought to be a carrier of vitamin A.

It is interesting to note that β-lactoglobulin is not present in human milk. α-Lactalbumin plays a critical role in the synthesis of lactose in the mammary gland. Immunoglobulins play a role in the animal's immune system, but it is unknown if these functions are transferred to humans. Lactoferrin and transferrin play an important role in iron absorption and there is interest in using bovine milk as a commercial source of lactoferrin.

Milk Protein Physical Properties

The caseins in milk form complexes called micelles that are dispersed in the water phase of milk. The casein micelles consist of subunits of the different caseins (α-s1, α-s2 and β) held together by calcium phosphate bridges on the inside, surrounded by a layer of 6-casein which helps to stabilize the micelle in solution.

Casein micelles are spherical and are 0.04 to 0.3 µm in diameter, much smaller than fat globules which are approximately 1 µm in homogenized milk. The casein micelles are porous structures that allow the water phase to move freely in and out of the micelle. Casein micelles are stable but dynamic structures that do not settle out of solution.

They can be heated to boiling or cooled, and they can be dried and reconstituted without adverse effects. β-casein, along with some calcium phosphate, will migrate in and out of the micelle with changes in temperature, but this does not affect the nutritional properties of the protein and minerals. The whey proteins exist as individual units dissolved in the water phase of milk.

Deterioration of Milk Protein

Proteins can be degraded by enzyme action or by exposure to light. The predominant cause of protein degradation is through enzymes called proteases. Milk proteases come from several sources: the native milk, airborne bacterial contamination, bacteria that are added intentionally for fermentation, or somatic cells present in milk. The action of proteases can be desirable, as in the case of yogurt and cheese manufacture, so, for these processes, bacteria with desirable proteolytic properties are added to the milk. Undesirable degradation (proteolysis) results in milk with off-flavours and poor quality. The most important protease in milk for cheese manufacturing is plasmin because it causes proteolysis during ripening which leads to desirable flavours and texture in cheese.

Two amino acids in milk, methionine and cystine are sensitive to light and may be degraded with exposure to light. This results in an off-flavour in the milk and loss of nutritional quality for these 2 amino acids.

Influence of Heat Treatment on Milk Proteins

The caseins are stable to heat treatment. Typical high temperature short time (HTST) pasteurization conditions will not affect the functional and nutritional properties of the casein proteins. High temperature treatments can cause interactions between casein and whey proteins that affect the functional but not the nutritional properties. For example, at high temperatures, β-lactoglobulin can form a layer over the casein micelle that prevents curd formation in cheese.

The whey proteins are more sensitive to heat than the caseins. HTST pasteurization will not affect the nutritional and functional properties of the whey proteins. Higher heat treatments may cause denaturation of β-lactoglobulin, which is an advantage in the production of some foods (yogurt) and ingredients because of the ability of the proteins to bind more water. Denaturation causes a change in the physical structure of proteins, but generally does not affect the amino acid composition and thus the nutritional properties. Severe heat treatments such as ultra high pasteurization may cause some damage to heat sensitive amino acids and slightly decrease the nutritional content of the milk. The whey protein α-lactalbumin, however, is very heat stable.

Milk Processing

In rural areas, milk may be processed fresh or sour. The choice depends on available equipment, product demand and on the quantities of milk available for processing. In Africa, smallholder milk-processing systems use mostly sour milk. Allowing milk to ferment prior to processing has a number of advantages and processing sour milk will continue to be important in this sector. Where greater volumes of milk can be assembled, processing fresh milk gives more product options, allows greater throughput of milk and, in some instances, greater recovery of milk solids in product.

Because of differences between processing systems, each will be dealt with separately. The section on fresh-milk technology deals with techniques used for processing fresh milk in batches of up to 500 litres. Sour-milk technology is used for processing batches of up to 15 litres of accumulated sour milk. This will be described in the section on sour-milk technology.

Fresh Milk Technology

This section describes the manufacture of skim milk, cream, butter, butter oil, ghee, boiled-curd and pickled cheese varieties and fermented milks from fresh milk. The processing scale envisaged is 100 to 200 litres of milk per day. However, the processes described are suitable for batches of up to 500 litres per day.

Most of the equipment described can be fabricated locally. Equipment not available locally, such as a milk separator, has a cost advantage and quickly gives a good financial return in terms of increased efficiency. Hand-operated milk separators are durable and have a long life when properly maintained. Importation of such equipment is, therefore, advantageous.

The procedures given here are very precise. In many rural dairy processing plants, however, monitoring equipment may not be available and, although yields may be maximised by adhering to the prescribed procedures, all these products can be successfully made by approximating temperature, time, pH etc to the best of one's ability. It is particularly important in cheese-making to proceed when the curd is in a suitable condition. Therefore, times given are only approximate and the processor will, with experience, adopt methods suitable to his/her own environment.

Milk Separation

The fat fraction separates from the skim milk when milk is allowed to stand for 30 to 40 minutes. This is known a 'creaming'. The creaming process can be used to remove fat from milk in a more concentrated form. A number of methods are employed to separate cream from milk. An understanding of the creaming process is necessary to maximise the efficiency of the separation process.

Gravity Separation

Fat globules in milk are lighter than the plasma phase, and hence rise to form a cream layer. The rate of rise (V) of the individual fat globule can be estimated using Stokes' Law which defines the rate of settling of spherical particles in a liquid:

$$V = (r^2 (d_1 - d_2)g)/9ç$$

where r = radius of fat globules

d_1 = density of the liquid phase

d_2 = density of the sphere

g = acceleration due to gravity, and

ç = specific viscosity of the liquid phase

Particle r^2: As temperature increases, fat expands and therefore r^2 increases. Since the sedimentation velocity of the particle increases in proportion to the square of the particle diameter, a particle of radius 2 ($r^2 = 4$) will settle four times as fast as a particle of radius 1 ($r^2 = 1$). Thus, heating increases sedimentation velocity.

$d_1 - d_2$: Sedimentation rate increases as the difference between d_1 and d_2 increases. Between 20 and 50°C, milk fat expands faster than the liquid phase on heating. Therefore, the difference between d_1 and d_2 increases with increasing temperature.

g: Acceleration due to gravity is constant. This will be considered when discussing centrifugal separation.

ç: Serum viscosity decreases with increasing temperature. Calculation of the sedimentation velocity of a fat globule reveals that it rises very slowly, As shown in the equation, the velocity of rise is directly proportional to the square of the radius of the globule. Larger globules overtake smaller ones quickly. When a large globule comes into contact with a smaller globule the two join and rise together even faster, primarily because of their greater effective radius. As they rise they come in contact with other globules, forming clusters of considerable size. These clusters rise much faster than individual globules. However, they do not behave strictly in accordance with Stokes' Law because they have an irregular shape and contain some milk serum.

Factors affecting creaming: Cream layer volume is greatest in milk that has high fat content and relatively large fat globules, because such milk contains more large clusters. However, temperature and agitation affect creaming, irrespective of the fat content of the milk. Heating to above 60°C reduces creaming; milk that is heated to above 100°C retains very little creaming ability. Excessive agitation disrupts normal cluster formation, but creaming in cold milk may be increased by mild agitation since such treatment favours larger, loosely packed clusters.

Batch separation by gravity: Cream can be separated from milk by allowing the milk to stand in a setting pan in cool place. There are two main methods.

Shallow pan: Milk, preferably fresh from the cow, is poured into a shallow pan 40 to 60 cm in diameter and about 10 cm deep. The pan should be in a cool place. After 36 hours practically all of the fat capable of rising by this method will have come to the surface, and the cream is skimmed off with a spoon or ladle (Figure). The skim milk usually contains about 0.5 to 0.6% butterfat.

Deep-setting: Milk, preferably fresh from the cow, is poured into a deep can of small diameter. The can is placed in cold water and kept as cool as possible. After 24 hours the separation is usually as complete as it is possible to secure by this method. The skim milk is removed through a tap at the bottom of the can. Under optimum conditions, the fat content of the skim milk averages about 0.2 or 0.3 %. The pans should be rinsed with water immediately after use, scrubbed with hot water and scalded with boiling water.

Centrifugal Separation

Gravity separation is slow and inefficient. Centrifugal separation is quicker and more efficient, leaving less than 0.1% fat in the separated milk, compared with 0.5–0.6% after gravity separation.

The centrifugal separator was invented in 1897. By the turn of the century it had altered the dairy industry by making centralised dairy processing possible for the first time. It also allowed removal of cream and recovery of the skim milk in a fresh state.

The separation of cream from milk in the centrifugal separator is based on the fact that when liquids of different specific gravities revolve around the same centre at the same distance with the same angular velocity, a greater centrifugal force is exerted on the heavier liquid than on the lighter one. Milk can be regarded as two liquids of different specific gravities, the serum and the fat.

Milk enters the rapidly revolving bowl at the top, the middle or the bottom of the bowl (Figure). When the bowl is revolving rapidly the force of gravity is overcome by the centrifugal force, which is 5000 to 10 000 times greater than gravitational force. Every particle in the rotating vessel is subjected to a force which is determined by the distance of the particle from the axis of rotation and its angular velocity.

If we substitute centrifugal acceleration expressed as $r_1 ù^2$ (where r_1 is the radial distance of the particle from the centre of rotation and $ù^2$ is a measurement of the angular velocity) for acceleration due to gravity (g), we obtain:

$$V = (r^2(d_1 - d_2) \, r_1 ù^2)/9ç$$

Thus, sedimentation rate is affected by $r_1 \dot{u}^2$. In gravity separation, the acceleration due to gravity is constant. In centrifugal separation, the centrifugal force acting on the particle can be altered by altering the speed of rotation of the separator bowl.

In separation, milk is introduced into separation channels at the outer edge of the disc stack and flows inwards. On the way through the channels, solid impurities are separated from the milk and thrown back along the undersides of the discs to the periphery of the separator bowl, where they collect in the sediment space. As the milk passes along the full radial width of the discs, the time passage allows even small particles to be separated. The cream, i.e. fat globules, is less dense than the skim milk and therefore settles inwards in the channels towards the axis of rotation and passes to an axial outlet. The skim milk moves outwards to the space outside the disc stack and then through a channel between the top of the disc stack and the conical hood of the separator bowl.

Efficiency of separation is influenced by four factors: the speed of the bowl, residence time in the bowl, the density differential between the fat and liquid phase and the size of the fat globules.

Speed of the separator. Reducing the speed of the separator to 12 rpm less than the recommended speed results in high fat losses, with up to 12% of the fat present remaining in the skim milk.

Residence time in the separator: Overloading the separator reduces the time that the milk spends in the separator and consequently reduces skimming efficiency. However, operating the separator below capacity gives no special advantage—it does not increase the skimming efficiency appreciably but increases the time needed to separate a given quantity of milk.

Effect of temperature: Freshly drawn, uncooled milk is ideal for exhaustive skimming. Such milk is relatively fluid and the fat is still in the form of liquid butterfat. If the temperature of the milk falls below 22°C skimming efficiency is seriously reduced. Milk must therefore be heated to liquify the fat. Heating milk to 50°C gives the optimum skimming efficiency.

Effect of the position of the cream screw: The cream screw regulates the ratio of skim milk to cream. Most separators permit a rather wide range of fat content of cream (18–50%) without adversely affecting skimming efficiency. However, production of cream containing less than 18% or more than 50% fat results in less efficient separation.

Other factors that affect the skimming efficiency are:
- The quality of the milk: Milk in poor physical condition or which is curdy will not separate completely.
- Maintenance of the separator: A separator in poor mechanical condition will not separate milk efficiently.

When separation is complete the separator must be dismantled and cleaned thoroughly.

Hand Separator

In order to understand how centrifugal separation works, we shall follow the course of milk through a separator bowl. As milk flows into a rapidly revolving bowl it is acted upon by both gravity and the centrifugal force generated by rotation. The centrifugal force is 5000 to 10 000 times that of gravity, and the effect of gravity thus becomes negligible. Therefore, milk entering the bowl is thrown to the outer wall of the bowl rather than falling to the bottom. Milk serum has a higher specific gravity than fat and is thrown to the outer part of the bowl while the cream is forced towards the centre of the bowl.

Assembling the Bowl

1. Fit the milk distributor to the central feed shaft.
2. Fit the discs on top of each other on the central shaft.
3. Fit the cream screw disc.
4. Next, fit the rubber ring to the base of the bowl.
5. Put on the bowl shell, ensuring that it fits to the inside of the base.
6. Finally, screw the bowl nut on top.

Now the bowl is assembled and ready for use. The rest of the separator is essentially a set of gears so arranged as to permit the spindle, on which the bowl is carried, to be turned at high speed. The gears are normally enclosed in an oil-filled case.

The bowl is usually supported from the bottom and has two bearings; one to support its weight and the second to hold it upright. The upper bearing is usually fitted inside a steel spring so that it can keep the bowl upright even if the frame of the machine is not exactly level. The assembled bowl is lowered into the receptacle, making sure that the head of the spindle fits correctly into the hollow of the central feed shaft.

Chemistry of Dairy Products 99

Operation

1. When the bowl is set, fit the skim milk spout and the cream spout.
2. Fit the regulating chamber on top of the bowl.
3. Put the float in the regulating chamber.
4. Put the supply can in position, making sure that the tap is directly above and at the centre of the float.
5. Pour warm (body temperature) water into the supply can.
6. Turn the crank handle, increasing speed slowly until the operating speed is reached: This will be indicated on the handle or in the manufacturer's manual of operation. The bell on the crank handle will stop ringing when the correct speed is reached.
7. Open the tap and allow the warm water to flow into the bowl. This rinses and heats the bowl and allows a smooth flow of milk and increases separation efficiency.
8. Next, put warm milk (37 – 40°C) into the supply can. Repeat steps 6 and 7 above and collect the skim milk and cream separately.
9. When all the milk is used up and the flow of cream stops, pour about 3 litres of the separated milk into the supply can to recover residual cream trapped between the discs.
10. Continue turning the crank handle and flush the separator with warm water.

Cleaning the separator: Many of the impurities in the milk collect as slime on the wall of the separator bowl. This slime contains remnants of milk, skim milk and cream, all of which will decompose and ferment unless removed promptly. If not washed and freed from all impurities the separator bowl becomes a source of microbial contamination. Skimming efficiency is also reduced when the separator bowl and discs are dirty. Milk deposits on the separator can cause corrosion.

Washing the separator: After flushing the separator with warm skim milk, the bowl should be flushed with clean water until the discharge from the skim milk spout is clean. This removes any residual milk solids and makes subsequent cleaning of the bowl easier. The bowl should then be dismantled. Wash all. parts of the bowl, bowl cover, discharge spouts, float supply tank and buckets with a brush, hot water and detergent. Rinse with scalding water. Allow the parts to drain in a clean place protected from dust and flies. This process should be followed after each separation.

Cream Screw Adjustment

The cream screw should be adjusted so that the fat content of the cream is about 33%. Producing excessively thin cream reduces the amount of separated milk available for other uses and increases the volume of cream to be handled. Low-fat cream is also more difficult to churn efficiently.

Cream containing more than 45% fat clogs the separator and causes excessive loss of fat in skim milk. Cream of abnormally high fat content also gives butter a greasy body due to lack of milk SNF. When adjusting the cream screw it is important to remember that it is very sensitive; a quarter turn of the screw is sufficient to change the percentage fat in the cream appreciably.

The fat content of whole milk influences the fat content of cream and this must be considered when adjusting the cream screw. For example, if the cream screw is set to separate milk at a ratio of 85 parts of separated milk to 15 parts of cream then, with all other conditions constant and assuming efficient separation, milk of 3% fat produces cream of 20% fat whereas milk of 4.5% fat produces cream of 30% fat. The fat content of the cream can be calculated using the following equation:

$$Fc = (Wm \times Fm)/Wc$$

Wm = weight of milk	Fm = fat content of milk
Wc = weight of cream	Fc = fat content of cream
In the first example,	Fc = (100 × 3)/15 = 20
In the second example,	Fc = (100 × 4.5)/15 = 30

Therefore the setting of the cream screw depends on the fat content of the milk being separated. The milk should be mixed thoroughly prior to separation to ensure even distribution of cream in the milk.

Separator Maintenance

- The gears must be well lubricated. Follow the directions of the manufacturer.
- The level of the lubricant must be kept constant; observe the oil level through the sight glass.

- The bowl must be perfectly balanced.
- The bowl should be cleaned thoroughly immediately after use to ensure proper functioning of the separator and for hygiene.

Calculations

Once milk passes through a separator it is recovered in two fractions, the high-fat cream fraction and the low-fat skim milk.

Assuming negligible loss of fat in the separator, the amount of fat entering the separator with the whole milk will be collected at the other side of the separator in either the cream or the skim milk. Therefore, if we separate 200 kg of milk containing 4.5% butterfat, what weight of cream containing 30% butterfat can we expect?

Let W_m = weight of milk

F_m = fat content of the milk

W_c = weight of cream

F_c = fat content of the cream

W_s = weight of skim milk

Assuming that all of the fat present in the milk is recovered in the cream, then:

$W_m \times F_m = W_c \times F_c$

and $W_m - W_c = W_s$

and $W_m - W_s = W_c$

Since $W_m \times F_m = W_c \times F_c$

$(W_m \times F_m)/F_c = W_c$

Therefore $W_s = W_m - (W_m \times F_m)/F_c = W_c$

In this case: $W_s = 200 - (200 \times 4.5)/30 = 200 - 30 = 170$ kg

Since $W_c = W_m - W_s$

$W_c = 200 - 170 = 30$ kg

Percentage yield of skim milk:

$= W_s \times 100)/W_m = (170 \times 100)/200 = 85\%$

Percentage cream (%W_c) = %W_m − %W_s = 100 − 85 = 15%

If in practice we obtain only 28 kg of cream containing 30% butterfat, then (2 × 0.30) kg or 0.6 kg of butterfat has not been recovered in the cream.

Since it is assumed that there are no significant losses of fat in the cream separator, the fat not recovered in the cream is lost in the skim milk.

Since 28 kg of cream was produced, and
Ws = Wm – Wc
then Ws = 200 – 28 = 172 kg

Thus there is 0.6 kg of fat in 172 kg of skim milk. The fat percentage of the skim milk is therefore:

(0.6 × 100)/172= 0.35%*

* The skim milk contains 0.35% fat, which may be incorporated in cottage cheese. If the skim milk is consumed, no nutritional loss occurs, but a financial loss is incurred since the fat is more valuable if sold as butter than as cottage cheese or if it is consumed directly. The percentage of fat in milk and in cream influences Wc and Ws where the fat is recovered in the cream.

If   Fm = 3 %
     Fc = 30%
     Wm = 100

Then Wc = Wm × Fm/Fc
Wc = 100 × 3/30 = 10 kg
Ws = Wm – Wc = 100 – 10 = 90 kg
whereas if   Fm = 4%
     Fc = 30
     Wc = 100

Then Wc =100 × 4/30 = 13.3 kg
Ws = 100 – 13.3 =86.6 kg

Standardisation of Milk and Cream

If fine adjustment of the fat content of cream is required, or if the fat content of whole milk must be reduced to a given level, skim milk must be added. This process is known as standardisation.

The usual method of making standardisation calculations is the Pearson's Square technique. To make this calculation, draw a square and write the desired fat percentage in the standardised product at its centre and write the fat percentage of the materials to be mixed on the upper and lower left-hand corners. Subtract diagonally across the square the smaller from the larger figure and place the remainders on the diagonally opposite corners. The figures on the right-hand corners indicate the ratio in which the materials should be mixed to obtain the desired fat percentage.

Chemistry of Dairy Products

The value on the top right-hand corner relates to the material on the top left-hand corner and the figure on the bottom right relates to the material at the bottom left corner.

Example 1

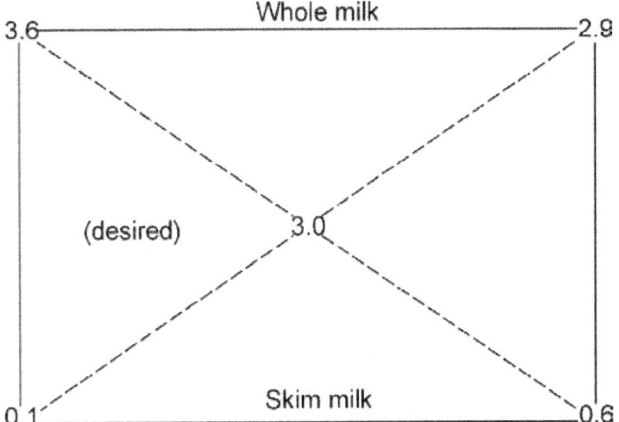

In this example, the fat content of whole milk is to be reduced to 3.0%, using skim milk produced from some of the whole milk. Using Pearson's Square, it can be seen that for every 2.9 litres of whole milk, 0.6 litres of skim milk must be added.

Example 2

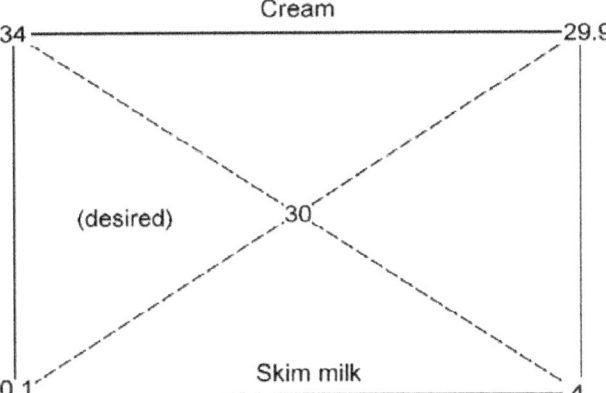

How much skim milk containing 0.1 % fat is needed to reduce the percentage fat in 200 kg of cream from 34% to 30%?

If 29.9 parts of cream require 4 parts of skim milk, 200 parts of cream require x parts of skim milk.

Weight of skim milk needed = x = (200 × 4)/29.9 = 26.75 kg

Example 3

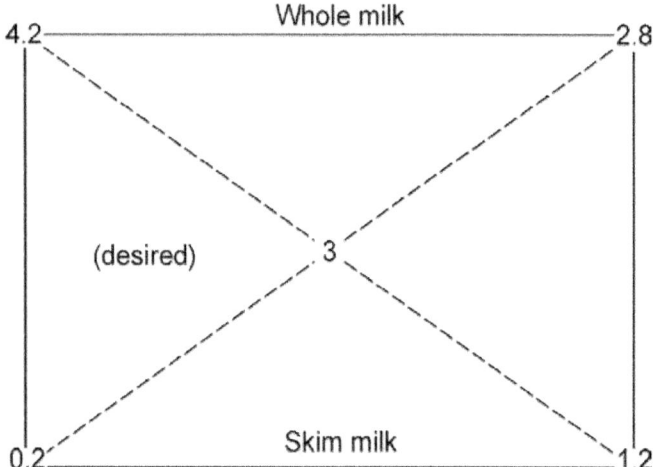

The fat content of 300 kg of whole milk must be reduced from 4.2% to 3% using skim milk containing 0.2% fat. Every 4.0 kg of the mixture will contain 2.8 kg of whole milk and 1.2 kg of skim milk.

If 2.8 kg of whole milk requires 1.2 kg skim milk, 300 kg of whole milk requires $(1.2 \times 300)/2.8 = 128.6$ kg of skim milk

Thus, 128.6 kg of skim milk (0.2% fat) must be added to 300 kg of whole milk (4.2% fat) to give 428.6 kg of milk containing 3% fat.

Example 4

The fat content of milk must be reduced from 4.5 to 3% prior to sale as liquid milk but skim milk for standardisation is not available.

In this case, we must calculate (a) what proportion of the milk must be separated to provide enough skim milk to standardise the remaining whole milk and (b) the expected yield of cream.

Assume that the fat content of 100 kg of milk containing 4.5% milk fat must be reduced to 3%. The amount of cream to be removed can be calculated as follows:

Let M = weight of milk to be standardised—in this example, 100 kg. Therefore M = 100

Fm = fat content of the original milk = 4.5

C = weight of cream

Fc = fat content of the cream = 35

SM = weight of standardised milk

Fsm = fat content of the standardised milk = 3.0

Since the milk is separated into cream and standardised milk
SM + C = M

(1) or SM + C = 100

There are no fat losses; therefore the weight of fat in the original milk will be equal to the weight of fat in the standardised milk and cream.

(Weight of fat in a product is the weight of product × % fat/100)

Therefore (SM × Fsm)/100 + (C × Fc)/100 = (M × Fm)/100

or (3 × SM)/100 + (35 × C)/100 = (100 × 4.5)/100

(2) or 0.03SM + 0.35C = 4.5

Equations (1) and (2) give two equations with two unknowns, so they can be solved as follows:

(1) SM + C = 100

(3) or 0.03SM + 0.03C = 3

Subtracting (3) from (2)

0.32 C = 1.5

C = 4.6875

= 4.7 corrected to one decimal place

The weight of cream is thus 4.7 kg.

Therefore, the weight of standardised milk is 95.3 kg.

Answer check

The original milk contained 4.5 kg of fat.

The cream contains (4.7 × 35)/100 =1.645 kg of fat

Therefore 4.5 − 1.645 = 2.855 kg of fat in the standardised milk.

The fat percentage of the standardised milk is

(2.855 × 100)/95.3 = 3%

The calculation can also be made using Pearson's Square. This is essentially a reverse standardisation, i.e. "how much cream containing 35% fat and milk containing 3% fat should be mixed to get milk containing 4.5% fat?" is mathematically the same as "how much cream containing 35% fat must be removed from milk containing 4.5% fat to standardise the milk to 3% fat content?"

1. Place the fat content of whole milk in the centre.
2. Place the fat content of cream on the top left-hand corner.

3. Place the desired fat content of the standardised milk on the bottom left-hand corner.
4. For every 32 parts of whole milk, there are 1.5 parts of cream to be removed and 30.5 parts of standardised milk.

Therefore Wc = (1.5)/32 × 100 = 4.6875 = 4.7

Wsm = Wm − Wc = 95.3

The Wsm and fat to be removed can be calculated in a number of ways. Whatever method is used to calculate the amount of cream to be removed, it is then necessary to calculate the amount of milk to be separated to achieve the desired reduction in fat content.

Wm × Fm = Wc × Fc

Therefore Wm × 4.5 = 4.7 × 35

and Wm = (4.7 × 35)/4.5 = 36.5

Therefore, 36.5 kg of milk are separated and the skim milk is then combined with the remaining whole milk. Standardisation such as this can be used to increase income from milk production as follows:

Assume liquid milk price of 70 cents/kg

Assume butter price of EB* 10/kg

Income from 100 kg of milk = EB 70

Income from 95.3 kg of milk = 66.71

Fat removed = Wc × Fc = 4.7 × 0.35 = 1.645

Expected butter yield = 1.9 kg

Income from butter = EB 19

Total income = EB 85.76

Margin = EB 15.76/ 100 kg of milk

*EB = Ethiopian birr (US$ 1 = EB 2.07)

Butter-making with Fresh Milk or Cream

Butterfat can be recovered from milk and converted to a number of products, the most common of which is butter. Butter is an emulsion of water in oil and has the following approximate composition:

Fat	80%
Moisture	16%
Salt	2%
Milk SNF	2%

In good butter the moisture is evenly dispersed throughout the butter in tiny droplets. In most dairying countries legislation defines the composition of butter; and butter makers conform to these standards insofar as is possible.

Butter can be made from either whole milk or cream. However, it is more efficient to make butter from cream than from whole milk.

Butter-making Theory

To make butter, milk or cream is agitated vigorously at a temperature at which the milk fat is partly sold and partly liquid. Churning efficiency is measured in terms of the time required to produce butter granules and by the loss of fat in the buttermilk. Efficiency is influenced markedly by churning temperature and by the acidity of the milk or cream.

In churning, cream is agitated in a partly filled chamber. This incorporates a large amount of air into the cream as bubbles. The resultant whipped cream occupies a larger volume than the original cream.

As agitation continues the whipped cream becomes coarser. Eventually the fat forms semi-solid butter granules, which rapidly increase in size and separate sharply from the liquid buttermilk. The remainder of the butter-making process consists of removing the buttermilk, kneading the butter granules into a coherent mass and adjusting the water and salt contents to the levels desired.

Theory of the Mechanism of Churning

In considering the mechanism of churning the following factors must be taken into account:
- The function of air;
- The release of stabilising material from the fat globule surface into the buttermilk;
- The differences in structure between butter and cream; and
- The temperature dependence of the process.

Air is thought to be necessary for the process, but some workers have demonstrated that milk or cream can be churned in the absence of air, although it takes longer.

About one half of the stabilising material is liberated into the buttermilk during churning.

It is thought that during churning the fat globule membrane substance spreads out over the surface of the air bubbles, partly denuding the globules of their protective layer, and that a liquid portion of the fat exudes from the globule and partly or entirely covers the globule, rendering it hydrophobic.

In this condition the globules tend to stick to the air bubbles. Free fat destabilises the foam, causing it to collapse. The partly destabilised globules clinging to the air bubbles thus collect in clusters cemented together by free fat. These clusters appear as butter grains.

Churning Cream

Cream prepared by gravitational or mechanical separation can be used. Good butter can be made in any type of churn provided it is clean and in good repair.

Churn Preparation

The churn is prepared by rinsing with cold water, scrubbing with salt and rinsing again with cold water. Alternatively, it can be scalded with water at 80°C. After the butter has been removed, the churn should be washed well with warm water, scalded with boiling water and left to air. When not in use wooden churns should be soaked occasionally with water.

A new churn should first be washed with tepid water, scrubbed with salt and then washed with hot water until the water comes away clear. A hot solution of salt should then be allowed to stand in the churn for a short time. After rinsing again with hot water the churn should be left to air for at least one day before being used.

Churning Temperature

The temperature of the cream during churning is of great importance. If too cool, butter formation is delayed and the grain is small and difficult to handle. If the temperature is too high, the yield of butter will be low, because a large proportion of the fat will remain in the buttermilk, and the butter will be spongy and of poor quality.

Cream should be churned at 10 –12°C in the hot season and at 14 –17°C in the cold season. The temperature may be raised by standing the vessel containing the cream in hot water, or may be lowered by standing the vessel in cold spring water for a few hours before the cream is churned. The churning temperature may also be adjusted by the water used to dilute the cream. In the hot season, the coldest water available should be used, preferably water that has been stored in a refrigerator.

The amount of cream to be churned should not exceed one half the volumetric capacity of the churn. An airtight churn should be ventilated frequently during the first 10 minutes of churning to release gases driven out of solution by the agitation. If butter is slow in forming, adding a little water which is warmer than the churning temperature, but never over 25°C, usually causes it to form more quickly. When the butter appears like wet maize meal, water (1 litre per 4 litres of cream) at 2°C below the churning temperature should be added. It may be necessary to add water a second time to maintain butter grains of the required size. Churning should cease when the butter grains are as large as small wheat grains.

Washing the Butter

When the desired grain size is obtained, the buttermilk is drained off and the butter washed several times in the churn. Each washing is done by adding only as much water as is needed to float the butter and then turning the churn a few times. The water is then drained off: As a general rule two washings will suffice but in very hot weather three may be necessary before the water comes away clear. In the hot season the coldest water available should be used for washing, and in the cold season water about 2 to 3°C colder than the churning temperature should be used.

Salting, Working and Packing the Butter

Equipment for working may consist of a butter worker or a tub or keeler. Good-quality spatulas are important, and a sieve and scoop facilitate the removal of butter from the churn. This equipment must be clean. The butter is spread on the worker, which has been soaked previously with water of the same temperature as the washing water. If salted butter is required, the butter should be salted before working at a rate of 16 g salt/kg or according to taste. The salt used should be dry and evenly ground and of the best quality available.

The butter is then either rolled out 8 to 10 times or ridged with the spatulas to remove excess moisture. If the butter is to be heavily salted, it must be worked more in proportion to the amount of salt used, as uneven distribution of the salt causes uneven colour. The butter should be worked until it seems dry and solid, but it must not be worked too much or it will become greasy and streaky.

The butter is then weighed and packed for storage. It should be packed in polythene-lined wooden or cardboard cartons and stored in a cool, dry place. The butter should be firm and of uniform colour.

Washing the Churn and Butter-making Equipment After Use

The churn and butter-making equipment should be washed as soon as possible, preferably while the wood is still damp.

Churn: Wash the inside of the churn thoroughly with hot water. Invert the churn with the lid on in order to clean the ventilator; this should be pressed a few times with the back of a scrubbing brush to allow water to pass through. Remove the rubber band from the lid and scrub the groove. Scald the inside of the churn with boiling water. This step is very important. Invert and leave to air. Dry the outside and treat steel parts with vaseline to prevent rusting. The rubber band should not be placed in boiling water; dipping in warm water is sufficient.

Butter worker/keeler: Place the sieve, scoop and spades on the butter worker or keeler. Pour hot water over all of them and scrub well to remove all traces of grease. Scald with boiling water and leave to air. Treat the steel part of the butter worker with vaseline to prevent rusting.

Storage of Butter

Surplus good-quality butter can be stored, but should contain more salt than usual—at least 30 g/kg. Low moisture content is desirable. The butter must be packed in clean containers, such as seasoned boxes or glazed crocks, and stored in a cold room or in a cold, airy place. If a box is used, it should be lined with good-quality polythene. The container should be filled to capacity from one churning. The more firmly butter is packed, the better; it may be covered with a layer of salt, but this is not essential. The container should be securely covered with a lid or a sheet of strong paper.

Overrun and Produce in Butter-making

An enterprise engaged in butter-making must be able to measure the efficiency of the process, i.e. by measuring the yield of butter from the butterfat purchased. First, the theoretical yield of butter has to be estimated. Butter contains an average of 80% butterfat. Thus, for every 80 kg of butterfat purchased 100 kg of butter should be produced, or for every 100 kg of butterfat purchased 125 kg of butter should be produced.

The difference between the number of kilograms of butterfat churned and the number of kilograms of butter made is known as the overrun. This difference is due to the fact that butter contains non-fatty constituents such as moisture, salt, curd and small amounts of lactic acid and ash in addition to butterfat.

The overrun is financially important to the dairy industry and constitutes the margin between the purchase price of butterfat and the sale price of butter. The dairy unit depends largely on overrun to cover manufacturing costs and to defray expenses incurred in the purchase of milk.

As stated above, the maximum legitimate overrun is 25%. In commercial operation, however, it is not possible to establish the degree of accuracy that is assumed in the calculation of theoretical overrun, and the actual overrun shows the difference between the amount of butter churned out and the amount of butterfat bought.

Overrun is affected by:
- Accuracy of weighing milk received.
- Accuracy of sampling and testing milk for fat.

Produce

Another method for estimating the efficiency of a process is to measure the number of litres of milk required per kilogram of butter produced.

For example, how many litres of milk containing 4% butterfat are required to make 1 kg of butter?

In 1 kg of butter there is 0.80 kg of butterfat.

In the milk we have 4 kg fat/ 100 kg or per 100 litres/1.032.

Therefore we have:

1 kg fat in $100/(1.032 \times 4) = 24.22$ litres

or 0.8 kg. fat in 19.38 litres

Therefore 19.38 litres of milk containing 4% fat will be required to make 1 kg of butter. Thus the efficiency of operation can also be checked by calculating output.

The fat content of the whole milk, skim milk and buttermilk should be checked daily. The moisture content of the butter should be checked for each batch. The accuracy of weighing scales and other measuring devices should be checked regularly.

Butter Quality

Butter quality can be discussed under two main headings:
- Compositional quality
- Organoleptic quality.

The compositional quality of butter can be further divided into two subsections:
- Chemical composition
- Bacteriological composition.

Compositional Quality

The chemical composition of butter is determined at the processing stage when the salt, moisture, curd and fat contents of the product are regulated. Once these parameters have been set there is little one can do to change them. The microbiological quality of butter is also determined during the production and processing stages.

Chemical composition affects butter yield, while butter of poor microbiological quality will deteriorate rapidly and become unacceptable to consumers. The butter may also contain pathogens. Cleanliness at all stages of production is, therefore, essential.

Organoleptic Quality

The organoleptic quality of butter can be described as the customer's reaction to its colour, texture and flavour. It has been said that the consumer tastes with his or her eyes, and it is true that a person's initial impression of a food will often determine whether or not he or she will buy it. It is important, therefore, to produce butter that has an even colour, clean flavour and close texture. It is also important that it be free from defects such as loose moisture. It should be packed attractively, both to attract customer attention and to retain its quality.

Butter produced carelessly and without the use of preservatives has a very short shelf life. Preservation of butter quality can assist the smallholder in two ways:
- The less perishable the product the longer the smallholder can retain it to obtain a good price.
- He or she can store the surplus made during the production season for consumption during the season in which he or she cannot produce butter.

The first step the producer can take to ensure a high-quality product is to make it in a clean, hygienic manner.

This results in fewer spoilage organisms being present in the butter. Another step is to take care in the handling and storage of the butter. The use of permitted preservatives is by far the most effective means of maintaining butter quality when used in conjunction with the above precautions. Salt—sodium chloride—is an excellent preservative, and salting butter to 3% extends its storage life: salted butter can be stored for up to 4 months without significant deterioration. A salt concentration in excess of 3% gives little advantage and can adversely affect the flavour of the butter.

Aside from the influence of salt on the flavour and keeping quality of the butter, adding salt is of economic importance as it increases overrun. Adding salt to butter disturbs the equilibrium of the emulsion (the butter). This, in turn, changes the character of the body and alters its colour. Unless the butter is subjected to sufficient working to regain the original equilibrium of the emulsion, it will tend to have a coarse, leaky body and uneven colour.

Salt is added to butter most commonly using the dry-salting method, in which dry salt is sprinkled evenly over the butter and worked in.

Butter must be adequately worked if it is to be stored for a long time. First, working distributes the salt uniformly in the moisture and this helps inhibit microbial growth. Secondly, it distributes the salt solution into many tiny droplets rather than fewer large ones. For a given level of microbial contamination, the microbes will be more isolated in small droplets and will have less of the butter's nutrients available to them for growth. After salting, the butter should be stored in a clean container, and the container sealed. It should then be stored in a cool, dark place.

Ghee, Butter Oil and Dry Butterfat

These products are almost entirely butterfat and contain practically no water or milk SNF. Ghee is made in eastern tropical countries, usually from buffalo milk. An identical product called *samn* is made in Sudan. Much of the typical flavour comes from the burned milk SNF remaining in the product. Butter oil or anhydrous milk fat is a refined product made by centrifuging melted butter or by separating milk fat from high-fat cream.

Ghee is a more convenient product than butter in the tropics because it keeps better under warm conditions. It has low moisture and milk SNF contents, which inhibits bacterial growth.

Milk or cream is churned as described in the sections dealing with churning of whole milk or cream. When enough butter has been accumulated it is placed in an iron pan and the water evaporated at a constant rate of boiling. Overheating must be avoided as it burns the curd and impairs the flavour. Eventually a scum forms on the surface: this can be removed using a perforated ladle. When all the moisture has evaporated the casein begins to char, indicating that the process is complete. The ghee can then be poured into an earthenware jar for storage.

A considerable amount of moisture and milk SNF can be removed prior to boiling by melting the butter in hot water (80°C) and separating the fat layer. The fat can be separated either by gravity or using a hand separator. The fat phase yields a product containing 1.5% moisture and little fat is lost in the aqueous phase.

Alternatively, the mixture can be allowed to settle in a vessel similar to that used in the deep-setting method for separating whole milk. Once the fat has solidified the aqueous phase is drained. The fat is then removed and heated to evaporate residual moisture. Products made using these methods exhibited excellent keeping qualities over a 5-month test period.

Cheese-making

Cheese is a concentrate of the milk constituents, mainly fat, casein and insoluble salts, together with water in which small amounts of soluble salts, lactose and albumin are found. To retain these constituents in concentrated form, milk is coagulated by direct acidification, by lactic acid produced by bacteria, by adding rennet, or a combination of acidification and addition of rennet.

Rennet Coagulation Theory

Rennet, a proteolytic enzyme extracted from the abomasum of suckling calves, was traditionally used for coagulating milk. Originally, the abomasum was itself immersed in milk. The extraction of rennet that could be stored as a liquid was the first step towards refining this procedure. This was followed by purification and concentration of the enzyme. The purified enzyme was originally called rennin, and is now called chymosin.

On weaning, the chymosin of the suckling calf is replaced by bovine pepsin. With the decrease in the practice of slaughtering calves, chymosin became scarce, resulting in a search for chymosin substitutes. Rennet is a general term currently used to describe a variety of enzymes of animal, plant or microbial origin used to coagulate milk in cheese-making.

Rennet transforms liquid milk into a gel. While the process is not fully understood, rennet coagulation is thought to take place in two distinct phases, the first of which is regarded as being enzymatic, the second non-enzymatic. The first, or primary phase, can be illustrated as:

$$\text{Casein} \xrightarrow[\text{rennet}]{\text{water}} \text{para casein} + \text{glycomacropeptide}$$

Since k-casein stabilises the other caseins and its hydrolysis leads to the coagulation of the casein fraction, the primary phase can also be expressed as:

$$\text{8-casein} \xrightarrow[\text{rennet}]{\text{water}} \underset{\text{(insoluble)}}{\text{para 8-casein}} + \underset{\text{(soluble)}}{\text{glycomacropeptide}}$$

The effect of milk coagulants on the other caseins is thought to be negligible at this stage.

The second, or secondary, phase is the non-enzymatic precipitation of para casein by calcium ions. Para casein, in association with the calcium ions, is thought to produce a lattice structure throughout the milk. This traps the fat and whey is gradually exuded. The coagulum then contracts, a process known as syneresis. This is accelerated by increasing the temperature and reducing pH to as low as pH 4.6.

Rennet also has a tertiary action on milk proteins. This occurs during cheese ripening, during which rennet hydrolyses milk proteins. If the desired hydrolysis is not obtained, the cheese becomes bitter. While a wide variety of proteolytic enzymes coagulate milk, the tertiary action of many of these on milk proteins causes undesirable flavours in cheese, which limits the range of coagulants that can be used.

Cheese Varieties

Many cheese varieties are manufactured around the world but they are all broadly classified by hardness (i.e. very hard, hard, semi-soft and soft) according to their moisture content.

Cheese is usually made from cows milk, although several varieties are made from the milk of goats, sheep or horses. Flow diagrams for the manufacture of the varieties discussed.

Queso Blanco (White Cheese)

Queso blanco is a Latin-American fresh, white cheese. It is usually made from milk containing 3% fat, using an organic acid, without starter or rennet.

Procedure

1. Take fresh whole milk and determine its fat content. If the fat content is higher than 3%, standardise using skim milk.
2. Transfer the standardised milk to a cheese vat, preferably a double-jacketed standard cheese vat, and heat to 82°C.
3. While the milk is being heated measure out lemon juice of pH about 2.5 in a measuring jar. About 3 ml of lemon juice should be added per 100 ml of milk.
4. Dilute the lemon juice with an equal amount of clean, fresh water.
5. When the milk temperature reaches 82°C, add the diluted lemon juice carefully and uniformly while stirring. For even distribution of the juice, add in three separate amounts.
6. The curd precipitates almost immediately. Continue to stir for 3 minutes after adding the juice, then allow the curd to settle for 15 minutes
7. Drain the whey through a metal sieve or cheese cloth.
8. While draining the whey, stir the curd to prevent excess matting.
9. Distribute a total of about 3.5 to 5 kg of salt to 100 kg curd, in three applications.
10. Prepare a cylindrical or square hoop by lining with cheese cloth and scoop the salted curds into it.
11. Press the curd overnight at room temperature.
12. Remove the pressed cheese and cut into blocks of 0.5 or 1 kg.

Queso blanco is made without starter or rennet. A variety of acidulants can be used for its manufacture. Heating the milk to 82°C pasteurises the milk and denatures the whey proteins, so that they are recovered with the curd. This increases cheese yield. The cheese has good keeping quality and is thus suitable for manufacture in rural areas.

Halloumi

Halloumi is the curd, formed by coagulating whole milk using rennet or similar enzymes, from which part of the moisture (whey) has been removed by cutting (bleeding), warming and pressing.

Procedure

1. Heat the milk to 32–35°C.
2. Add rennet or a similar enzyme according to the manufacturer's directions, while stirring the milk.
3. Hold the milk at 32–35°C until the curd sets.
4. Check for setting of the curd by applying pressure to the edge of the milk where it comes in contact with the vat, using a spatula or a knife with a round tip. If the curd is set it comes away clean from the wall of the vat.
5. After coagulation, the curd is cut into 3–5 mm cubes using vertical and horizontal knives.
6. Hold the curd in whey for about 20 minutes, stirring gently and continuously, and then allow it to settle.
7. Drain the whey and scoop out the curd into a hoop lined with cheese cloth. Press the curd.
8. While the curd is in the press, heat the whey to about 80–90°C. This precipitates the whey proteins, which can then be removed and pressed to make a whey cheese.
9. Take out the pressed curd, cut it into pieces of 10 × 10 × 3 cm and heat at about 80°C in hot whey. Continue heating until the pieces of curd float on the surface of the whey and become soft and elastic.
10. Remove the pieces of curd when still warm and either press in the hands, folded or unfolded, and rub in a little dry salt mixed with dried leaves of *Mentha viridis* (spearmint).
11. When the pieces are cold, put them in containers filled with cool, boiled whey brine and store in a cool place to ripen for about 30 days.
12. After ripening put in an airtight container and store in a refrigerator at less than 12°C. The cheese will keep for several months under these conditions. Halloumi cheese is best after 40 days but can also be consumed just after manufacture.

Domiati–Gybna Beyda

Known as Domiati in Egypt and Gybna beyda in Sudan, this is a hard, white cheese.

Procedure

1. Heat fresh milk to 35°C and add enough salt to give 7 to 10% salt solution in the milk.
2. Add enough rennet to coagulate the milk in 4 to 6 hours.
3. Once set, transfer the coagulum to wooden moulds lined with muslin.
4. Allow the whey to drain overnight.
5. On the following day, pack the cheese in tins and fill the interspaces with whey.
6. Seal the tins by soldering.

Notes:

1 and 2. In some areas rennet is added before salting. In this procedure, salt is not added until a coagulum has formed. If salt is added before rennet it is not advisable to add more rennet to shorten the coagulation time, as this reduces the quality of the cheese.

Whey expulsion continues during storage and the cheese hardens. Expected yield: 1 kg of cheese from 7 kg of milk (15%).

Feta

This is a brine-pickled cheese. It can be made from milk of cows, sheep or goats. Feta can be made without starter and can also be made from standardised milk. The procedure described here is for the manufacture of a feta-type cheese without starter or additives.

Procedure

- Standardise the milk to 3% fat, heat to about 32°C and allow to ripen for one hour before adding rennet.
- Add commercial rennet at the rate of 25 ml/ 100 litres of milk. Leave the milk until a firm clot has formed—this usually takes 40 to 50 minutes.
- Cut the curd into 2-to 3-cm cubes to facilitate whey drainage. Allow 15 minutes for the whey to separate. Stir intermittently during this time.
- Allow the curds to settle and decant the supernatant whey.

- Transfer the curds and some whey to cheese moulds lined with muslin. Place the lid on the mould and invert at half-hourly intervals in the first few hours to facilitate whey drainage.
- Allow the curd to settle overnight.
- On the following day, cut the curd mass into blocks of suitable size and sprinkle them with salt.
- Place the salted blocks in a 15% brine solution to give 6–8% salt in the cheese at equilibrium.

The high salt concentration retards bacterial activity. However, air should be excluded from the brining container to prevent the growth of moulds.

Feta cheese can be eaten after a few days or can be stored for long periods in the brine, provided that air is excluded. The cheese develops a soft, crumbly texture during ripening.

Expected yield: 1 kg of cheese from 9 kg of milk (11 %).

Cheese Yield

In cheese-making, the milk fat and casein are recovered with some moisture. The yield of cheese can be expressed in kilograms of cheese obtained per 100 kilograms of milk processed. Cheese yield is influenced by milk composition, the moisture content of the final cheese and the degree of recovery of fat and protein in the curd during cheese-making.

Milk low in total solids will give a low cheese yield, while milk high in total solids will give a high cheese yield. In order to predict the theoretical yield of cheese, the fat and casein content of the milk must be known. Because of difficulties encountered in estimating casein content, the following formula is often used to estimate cheese yield:

$(2.3 \times \text{fat \%}) + 1.4 = \text{cheese yield (kg/ 100 kg milk)}$

Therefore, with milk containing 4% fat the expected yield would be:

$(2.3 \times 4) + 1.4 = 10.6 \text{ kg/ 100 kg milk}$

This formula gives an estimate of cheese yield and is applied most often to Cheddar cheese. It is useful as an immediate check on efficiency, but a universal yield factor for cheese varieties is unrealistic.

If the yield of cheese is less than expected, the following checks should be made:

- Weigh and record milk received.
- Sample and analyse milk received.
- Weigh, store and record cheese made.
- Sample and analyse whey.

The fat content of whey should be analysed for each batch of cheese made.

In estimating the profitability of cheese-making enterprises, an average annual yield of 9.5%, i.e. 9.5 kg of cheese per 100 kg of milk, is used. Milk standardisation may be used to increase cheese yield, particularly with high-fat milk. Standardisation also gives a good return for skim milk. However, over-standardising results in coarse-textured cheese with poor flavour.

High moisture content increases cheese yield, but reduces keeping quality. Cheese loses moisture during storage if it is not properly wrapped, thus reducing cheese yield. Waxing reduces moisture loss, as does storing the cheese in brine.

Milk Fermentations

Raw milk produced under normal conditions develops acidity. It has long been recognised that highly acid milk does not putrefy. Therefore, allowing milk to develop acidity naturally preserves the other milk constituents.

Bacteria in milk are responsible for acid development. They produce acid by the anaerobic breakdown of milk carbohydrate—lactose—to lactic acid and other organic acids. Anaerobic breakdown of carbohydrate to organic acids or alcohols is called fermentation.

Pyruvic acid formation is an intermediate step common to most carbohydrate fermentations:

$$C_6H_{12}O_6 \longrightarrow 2\ CH_3.CO.COOH$$

However, fermentations are usually described by an identifiable end product such as lactic acid or ethyl alcohol and carbon dioxide.

A number of sugar fermentations are recognised in milk. They can be either homofermentative, with one end product, or heterofermentative, with more than one end product.

Organisms responsible:

1. Streptococci and Lactobacilli.
2. Propionibacteria.

Chemistry of Dairy Products 121

3. Yeasts – Candida and Torula.
4. Coliform bacteria.

- The lactic acid fermentation is the most important one in milk and is central to many processes.
- Propionic fermentation is a mixed-acid fermentation and is used in the manufacture of Swiss cheese varieties.
- Alcohol fermentation can be used to prepare certain fermented milks and also to make ethyl alcohol from whey.
- The coliform gassy fermentation is an example of a spoilage fermentation. Large numbers of coliform bacteria in milk indicates poor hygiene. The coliform gassy fermentation disrupts lactic acid fermentation, and also causes spoilage in cheese.

The factors that affect microbial growth also affect milk fermentation. Fermentation rates will generally parallel the microbial growth curve up to the stationary phase. The type of fermentation obtained will depend on the numbers and types of bacteria in the milk, storage temperature and the presence or absence of inhibitory substances. The desired fermentations can be obtained by temperature manipulation or by adding a selected culture of microorganisms—starter—to pasteurised or sterilised milk.

In smallholder milk processing, traces of milk from previous batches are often used to provide 'starter' for subsequent batches. Other sources include the container and additives such as cereal grains. The fermentation will be established once the organisms dominate the medium and will continue until either the substrate is depleted or the end product accumulates.

In milk, accumulation of end product usually arrests the fermentation. For example, accumulation of lactic acid reduces milk pH to below 4.5, which inhibits the growth of most microorganisms, including lactic-acid producers. The fermentation then slows and finally stops.

Fermented milks are wholesome foods and many have medicinal properties attributed to them.

Fermented Milks

The types of fermented milk discussed here are those made by controlled fermentation. This is achieved by establishing the desired microorganisms in the milk and by maintaining the milk at a temperature favourable to the fermentative organism. A variety of fermented milks are made, each dithering markedly from the other. However, a number of steps are common to each manufacturing process, and these are outlined.

Standardisation

Occasionally some fat is removed or milk SNF added. In some instances, the removal of moisture during heating increases the proportion of solids in the final product.

Heating

Milk is heated to kill pathogens and spoilage organisms and to provide a cleaner medium in which the desired microorganisms can be established. Heating also removes air from the milk, resulting in a more favourable environment for the fermentative organisms, and denatures the whey proteins, which increases the viscosity of the product.

After heating, the milk must be cooled before it is inoculated with starter, otherwise the starter organisms will also be killed.

Inoculation with Starter

Starter is the term used to describe the microbial culture that is used to produce the desired fermentation and to flavour the product. When preparing the starter, care must be taken to avoid contamination with other microorganisms. Companies that supply starter cultures detail the precautions necessary. Care should also be taken to avoid contamination when inoculating the milk with starter.

Incubation

After inoculation the milk is incubated at the optimum temperature for the growth of the starter organism. Incubation is continued until the fermentation is complete, at which time the product is cooled. Additives may be added at this stage and the product packed.

Preparation of the Fermentation Vessel: The fermentation vessel is first washed to remove visible dirt. It is then dried and smoked by putting burning embers of *Olea africana,* wattle or acacia into the vessel and closing the lid. The vessel is then shaken vigorously and the lid opened to release the smoke. This procedure is repeated until the inside of the vessel is hot. Smoking flavours the product and is also thought to control the fermentation by retarding bacterial growth. While it is known that smoke contains compounds that retard

Chemistry of Dairy Products

bacterial growth, the precise effects of smoking on fermentation have not been investigated.

Once smoking is complete the vessel may be cleaned with a cloth to remove charcoal particles. However, in some areas the charcoal particles are retained to add colour to the product.

Milk Treatment

In some processes the milk is boiled prior to fermentation. It is then allowed to cool and the surface cream removed. In other processes the milk is not given any prefermentation treatment.

Fermentation

The milk is placed in the smoked vessel and allowed to ferment slowly in a cool place at a temperature of about 16–18°C. The fermentation is almost complete after 2 days, but may be continued for a further 2 days, by which time the flavour is fully developed. The milk must ferment at low temperature, otherwise fermentation is too vigorous, with much wheying off and gas production.

The product has a storage stability of 15 to 20 days.

Concentrated Fermented Milks

Concentrated fermented milks are prepared by removing whey from fermented milk and adding fresh milk to the residual milk constituents. The fermentation vessel is prepared as for fermented milk. The milk is allowed to ferment in a cool place for up to 7 days, during which milk may be added daily.

After 7 days a coagulum has formed and the clear whey is removed. Fresh milk is then added and, following further fermentation, whey is again removed. In this way the casein and fat are gradually concentrated in a product of extended keeping quality. The actual degree of concentration depends on the amount of whey removed and of fresh milk added.

Sour-milk Technology

Smallholder milk processing is based on sour milk. This is due to a number of reasons, including high ambient temperatures, small daily quantities of milk, consumer preference and increased keeping quality of sour milk.

Products made from sour milk include fermented milks, concentrated fermented milks, butter, ghee, cottage cheese and whey.

Other products made are cheese and products made by mixing fermented milk with boiled cereals.

The equipment required for processing sour milk is simple and is all available locally. Milk vessels can be made from clay, gourds and wood, and can be woven from fibre, such as the *gorfu* container used by the Borana pastoralists in Ethiopia.

Butter-Making from Sour Whole Milk

This is a very important process in many parts of Africa. Smallholders produce 1 to 4 litres of milk per day for processing. Under normal storage conditions the milk becomes sour in 4 to 5 hours. The souring of milk has a number of advantages. It retards the growth of undesirable microorganisms, such as pathogens and putrefactive bacteria, and makes the milk easier to churn.

Milk for churning is accumulated over several days by adding fresh milk to the milk already accumulated. The churn holds about 20 litres and the amount of milk churned ranges from 4 to 10 litres. The milk is normally accumulated over 2 or more days. Butter is made by agitating the milk until butter grains form. The churn is then rotated slowly until the fat coalesces into a continuous mass. The butter thus formed is taken from the churn and kneaded in cold water.

The milk is usually agitated by placing the churn on a mat on the floor and rolling it to and from. It can also be agitated by shaking the churn on the lap or hung from a tripod.

A number of factors influence churning time and recovery of butterfat as butter:

- Milk acidity
- Churning temperature
- Degree of agitation, and
- Extent of filling the churn.

Effect of acidity: Fresh milk is difficult to churn: churning time is long and recovery of butterfat is poor. Milk containing at least 0.6% lactic acid is easier to churn. Acidity higher than 0.6% does not significantly influence churning time or fat recovery.

Effect of temperature: Sour milk is normally churned at between 15 and 26°C, depending on environmental temperature. At low temperatures churning time is long; butter-grain formation can take 5 hours or longer. As churning temperature increases churning time decreases. This becomes marked at temperatures above 20°C, but as little as 60% of the butterfat may be recovered as butter at 26°C. Control of temperature is therefore critical.

It is difficult to isolate the effects of temperature and acidity on churning efficiency because while the milk is ripening it is also cooling and the fat is crystallising. Direct acidification of fresh milk increases butter yield, but allowing milk to develop acidity during a ripening period of 2 to 3 days allows considerable fat crystallisation.

Degree of agitation: Increasing agitation reduces churning time. Fitting an agitator to a traditional churn reduces churning time and increases butter yield. The percentage of fat recovered as butter is increased, with as little as 0.2% fat remaining in the buttermilk. However, the process is very temperature-dependent and churning at temperatures above 20°C results in short churning times with poor recovery of fat. The optimum churning temperature is between 17 and 19°C.

Extent of filling the churn: Churns should be filled to between a third and half their volumetric capacity. Filling to more than half the volumetric capacity increases churning time considerably but does not reduce fat recovery.

Thus, when churning whole milk, the following conditions should be adhered to:
- Milk acidity should be greater than 0.6%.
- The temperature should be regulated to about 18°C.
- Internal agitation should be used to reduce churning time and increase fat recovery.
- The churn should not be filled to more than half its volumetric capacity.

Once the fat has been recovered, the soured skim milk contains casein, whey proteins, milk salts, lactic acid, lactose, the unrecovered fat and some fat-globule-membrane constituents.

Defatted milk is suitable, and is often used, for direct consumption. It is also used to inoculate fresh milk to encourage acid development.

Cottage Cheese

The casein and some of the unrecovered fat in skim milk can be heat-precipitated as cottage cheese, known in Ethiopia as *ayib*.

The defatted milk is heated to about 50°C until a distinct curd mass forms. It is then allowed to cool gradually and the curd is ladled out. Alternatively, the curd can be recovered by filtering the cooled mixture through a muslin cloth. This facilitates more complete recovery of the curd and also allows more effective moisture removal. Temperature can be varied between 40 and 70°C without markedly affecting product composition and yield. Heat treatments between 70 and 90°C do not appear to affect yield but give the product a cooked flavour.

The whey contains about 0.75% protein, indicating near-complete recovery of casein. Whey can be consumed by humans or fed to animals.

The cottage cheese comprises 79.5% water, 14.7% protein, 1.8% fat, 0.9% ash and 3.1 % soluble milk constituents. It has a short shelf-life because of its high moisture content. Shelf-life can be increased by adding salt or by reducing the moisture content of the cheese. Storing the product in an air-tight container also extends storage life.

Equipment: Skim milk can be heated in any suitably sized vessel that is able to withstand heat. Heating can be direct or indirect. A ladle or muslin cloth can be used for product recovery.

Expected yield: The yield depends on milk composition and on the moisture content of the product, but should be at least 1 kg of cottage cheese from 8 litres of milk (12.5%).

5

Fluid Milk Processing

Beverage Milks

The production of beverage milks combines the unit operations of clarification, separation (for the production of lower fat milks), pasteurization, and homogenization. The process is simple, as indicated in the flow chart. While the fat content of most raw milk is 4% or higher, the fat content in most beverage milks has been reduced to 3.4%. Lower fat alternatives, such as 2% fat, 1% fat, or skim milk (<0.1% fat) or also available in most markets. These products are either produced by partially skimming the whole milk, or by completely skimming it and then adding an appropriate amount of cream back to achieve the desired final fat content.

Vitamins may be added to both full fat and reduced fat milks. Vitamins A and D (the fat soluble ones) are often supplemented in the form of a water soluble emulsion to offset that quantity lost in the fat separation process.

Creams

During the separation of whole milk, two streams are produced: the fat-depleted stream, which produces the beverage milks as described above or skim milk for evaporation and possibly for subsequent drying, and the fat-rich stream, the cream. This usually comes off the separator with fat contents in the 35-45% range. Cream is used for further processing in the dairy industry for the production of ice cream or butter, or can be sold to other food processing industries. These industrial products normally have higher fat contents than creams for retail sale, normally in the range of 45-50% fat. A product known as "plastic" cream can be produced from certain types of milk separators.

This product has a fat content approaching 80% fat, but it remains as an oil-in-water emulsion (the fat is still in the form of globules and the skim milk is the continuous phase of the emulsion), unlike butter which also has a fat content of 80% but which has been churned so that the fat occupies the continuous phase and the skim milk is dispersed throughout in the form of tiny droplets (a water-in-oil emulsion).

For retail cream products, the fat is normally standardized to 35% (heavy cream for whipping), 18% or 10% (cream for coffee or cereal). Higher fat creams have also been produced for retail sale, a product known as double cream has a fat content of 55% and is quite thick. Creams for packaging and sale in the retail market must be pasteurized to ensure freedom from pathogenic bacteria. Whipping cream is not normally homogenized, as the high fat content will lead to extensive fat globule aggregation and clustering, which leads to excessive viscosity and a loss of whipping ability. This phenomena has been used, however, to produce a spoonable cream product to be used as a dessert topping. Lower fat creams (10% or 18%) can be homogenized, usually at lower pressure than whole milk.

Recombined Milk

Beverage milks can also be prepared by recombining skim milk powder and butter with water. This is often done in countries where there is not enough milk production to meet the demand for beverage milk consumption. The concept is simple. Skim milk powder is dispersed in water and allowed to hydrate. Butter is then emulsified into this mixture by either blending melted butter into the liquid mixture while hot, or by dispersing solid butter into the liquid through a high shear blender device.

In some cases, a non-dairy fat source may also be used. The recombined milk product is then pasteurized, homogenized and packaged as in regular milk production. The final composition is similar to that of whole milk, approximately 9% milk solids-not-fat, and either 2% or 3.4% fat. The water source must be of excellent quality. The milk powder used for recombining must be of high quality and good flavour. Care must be taken to ensure adequate blending of the ingredients to prevent aggregation or lumping of the powder. Its dispersal in water is the key to success.

Chocolate Milk

An industry standard for the production of chocolate milk consists of:

- 93% milk
- 6.3% sugar
- 0.65% cocoa powder
- 0.05% carrageenan.

The final product is usually standardized to either 2% fat or 1% fat (meaning, 2.15% or 1.1% fat in the milk before addition of other ingredients). The sugar, cocoa powder and carrageenan are dry blended, and added to cold milk with vigourous agitation, and then pasteurized.

Concentrated and Dried Dairy Products

Fluid milk contains approximately 88% water. Concentrated milk products are obtained through partial water removal. Dried dairy products have even greater amounts of water removed to usually less than 4%. The benefits of both these processes include an increased shelf-life, convenience, product flexibility, decreased transportation costs, and storage.

The following products will be discussed here:
- Concentrated Dairy Products
- Evaporated Skim or Whole Milk.

After the raw milk is clarified and standardized, it is given a preheating treatment of 93-100° C for 10 to 25 min or 115-128° C for 1 to 6 min.. There are several benefits to this treatment:
- increases the concentrated milk stability during sterilization; decreases the chance of coagulation taking place during storage
- decreases the initial microbial load
- modifies the viscosity of the final product
- milk enters the evaporator already hot.

Milk is then concentrated at low temperatures by vacuum evaporation. This process is based on the physical law that the boiling point of a liquid is lowered when the liquid is exposed to a pressure below atmospheric pressure. In this case, the boiling point is lowered to approximately 40-45° C. This results in little to no cooked flavour. The milk is concentrated to 30-40% total solids. The evaporated milk is then homogenized to improve the milkfat emulsion stability. There are other benefits particular to this type of product:
- increased white colour
- increased viscosity
- decreased coagulation ability.

A second standardization is done at this time to ensure the proper salt balance is present. The ability of milk to withstand intensive heat treatment depends to a great degree on its salt balance.

The product at this point is quite perishable. The fat is easily oxidized and the microbial load, although decreased, is still a threat. The evaporated milk at this stage is often shipped by the tanker for use in other products.

In order to extend the shelf life, evaporated milk can be packaged in cans and then sterilized in an autoclave. Continuous flow sterilization followed by packaging under aseptic conditions is also done. While the sterilization process produces a light brown colouration, the product can be successfully stored for up to a year.

Sweetened Condensed Milk

Where evaporated milk uses sterilization to extend its shelf-life, sweetened condensed milk has an extended shelf-life due to the addition of sugar. Sucrose, in the form of crystals or solution, increases the *osmotic pressure* of the liquid. This in turn, prevents the growth of microorganisms.

The only real heat treatment (85-90° C for several seconds) this product recieves is after the raw milk has been clarified and standardized. The benefits of this treatment include totally destroying osmophilic and thermophilic microorganisms, inactivating lipases and proteases, decreases fat separation and inhibits oxidative changes. Unfortunately it also affects the final product viscosity and may promote the defect age gelation.

The milk is evaporated in a manner similar to the evaporated milk. Although sugar may be added before evaporation, post evaporation addition is recommended to avoid undesirable viscosity changes during storage.

Enough sugar is added so that the final concentration of sugar is approximately 45%.

The sweetened evaporated milk is then cooled and lactose crystallization is induced. The milk is inoculated, or seeded, with powdered lactose crystals, then rapidly cooled while being agitated. The lactose can crystalize without the seeding but there is the danger of forming crystals that are too large. This would result in a texture defect similar in ice cream called sandiness, which affects the mouthfeel. By seeding, the number of crystals increases and the size of those crystals decreases.

The product is packaged in smaller containers, such as cans, for retail sales and bulk containers for industrial sales.

Condensed Buttermilk

Buttermilk is a by-product of the butter industry. It can be evaporated on its own or it can be blended with skimmilk and dried to produce skimmilk powder. This blended product may oxidise readily due to the higher fat content. Condensed buttermilk is perishable and, therefore, the supply must be fresh and it must be stored cool.

Condensed Whey

In the process of cheesemaking, there is alot of whey that needs to be disposed of. One of the ways of utilizing cheesewhey is to condense it. The whey contains fat, lactose, β-lactoglobulin, alpha-lactalbumin, and water. The fat is generally removed by centrifugation and churned as whey cream or used in ice cream. Evaporation is the first step in producing whey powder.

Dried Dairy Products

Milk Powder

Milk used in the production of milk powders is first clarified, standardized and then given a heat treatment. This heat treatment is usually more severe than that required for pasteurization. Besides destroying all the pathogenic and most of the spoilage microorganisms, it also inactivates the enzyme lipase which could cause lipolysis during storage.

The milk is then evaporated prior to drying for the following reasons:
- less occluded air and longer shelf life for the powder
- viscosity increase leads to larger powder particles
- less energy required to remove part of water by evaporation; more economical.

Homogenization may be applied to decrease the free fat content. Spray drying is the most used method for producing milk powders. After drying, the powder must be packaged in containers able to provide protection from moisture, air, light, etc. Whole milk powder can then be stored for long periods (up to about 6 months) of time at ambient temperatures.

Skim milk powder (SMP) processing is similar to that described above except for the following points:
1. contains less milkfat (0.05-0.10%)
2. heat treatment prior to evaporation can be more or less severe
3. homogenization not required
4. maximum shelf life extended to approximately 3 years.

Low-heat SMP is given a pasteurization heat treatment and is used in the production of cheese, baby foods etc. High-heat SMP requires a more intense heat treatment in addition to pasteurization. This product is used in the bakery industry, chocolate industry, and other foods where a high degree of protein denaturation is required.

Instant milk powder is produced by partially rehydrating the dried milk powder particles causing them to become sticky and agglomerate. The water is then removed by drying resulting in an increased amount of air incorporated between the powder particles.

Whey Powder

Whey is the by-product in the manufacturing of cheese and casein. Disposing of this whey has long been a problem. For environmental reasons it cannot be discharged into lakes and rivers; for economical reasons it is not desirable to simply dump it to waste treatment facilities. Converting whey into powder has led to a number products that it can be incorporated into.

It is most desirable, if and where possible, to use it for human food, as it contains a small but valuable protein component. It is also feasible to use it as animal feed. Between the pet food industry and animal feed mixers, hundred's of millions of pounds are sold every year. The feed industry may be the largest consumer of dried whey and whey products.

Whey powder is essentially produced by the same method as other milk powders. Reverse osmosis can be used to partially concentrate the whey prior to vacuum evaporation. Before the whey concentrate is spray dried, lactose crystallization is induced to decrease the hygroscopicity. This is accomplished by quick cooling in flash coolers after evaporation. Crystallization continues in agitated tanks for 4 to 24 h.

A fluidized bed may be used to produce large agglomerated particles with free-flowing, non-hygroscopic, no caking characteristics.

Whey Protein Concentrates

Both whey disposal problems and high-quality animal protein shortages have increased worldwide interest in whey protein concentrates. After clarification and pasteurization, the whey is cooled and held to stabilize the calcium phosphate complex, which later decreases membrane fouling.

The whey is commonly processed using ultrafiltration, although reverse osmosis, microfiltration, and demineralization methods can be used. During ultrafiltration, the low molecular weight compounds such as lactose, minerals, vitamins and nonprotein nitrogen are removed in the permeate while the proteins become concentrated in the retentate. After ultrafiltration, the retentate is pasteurized, may be evaporated, then dried. Drying, usually spray drying, is done at lower temperatures than for milk in order that large amounts of protein denaturation may be avoided.

Cheese

Traditionally, cheese was made as a way of preserving the nutrients of milk. In a simple definition, cheese is the fresh or ripened product obtained after coagulation and whey separation of milk, cream or partly skimmed milk, buttermilk or a mixture of these products. It is essentially the product of selective concentration of milk. Thousands of varieties of cheeses have evolved that are characteristic of various regions of the world.

Treatment of Milk for Cheesemaking

Like most dairy products, cheesemilk must first be clarified, separated and standardized. The milk may then be subjected to a sub-pasteurization treatment of 63-65° C for 15 to 16 sec. This thermization treatment results in a reduction of high initial bacteria counts before storage. It must be followed by proper pasteurization. While HTST pasteurization (72° C for 16 sec) is often used, an alternative heat treatment of 60° C for 16 sec may also be used. This less severe heat treatment is thought to result in a better final flavour cheese by preserving some of the natural flora. If used, the cheese must be stored for 60 days prior to sale, which is similar to the regulations for raw milk cheese.

Homogenization is not usually done for most cheesemilk. It disrupts the fat globules and increases the fat surface area where casein particles adsorb. This results in a soft, weak curd at renneting and increased hydrolytic rancidity.

Additives

The following may all be added to the cheese milk:
- Calcium choride
- nitrates
- colour
- hydrogen peroxide
- lipases.

Calcium choride is added to replace calcium redistributed during pasteurization. Milk coagulation by rennet during cheese making requires an optimum balance among ionic calcium and both soluble insoluble calcium phosphate salts. Because calcium phosphates have reverse solubility with respect to temperature, the heat treatment from pasteurization causes the equilibrium to shift towards insoluble forms and depletes both soluble calcium phosphates and ionic calcium. Near normal equilibrium is restored during 24-48 hours of cold storage, but cheese makers can't wait that long, so $CaCl_2$ is added to restore ionic calcium and improve rennetability. The calcium assists in coagulation and reduces the amount of rennet required.

Sodium or potassium nitrate is added to the milk to control the undesirable effects of *Clostridium tyrobutyricum* in cheeses such as Edam, Gouda, and Swiss.

Because milk colour varies from season to season, colour may added to standardize the colour of the cheese throughout the year. Annato, Beta-carotene, and paprika are used. The addition of hydrogen peroxide is sometimes used as an alternative treatment for full pateurization.

Lipases, normally present in raw milk, are inactivated during pasteurization. The addition of kid goat lipases are common to ensure proper flavour development through fat hydrolysis.

Inoculation and Milk Ripening

The basis of cheesemaking relies on the fermentation of lactose by lactic acid bacteria (LAB). LAB produce lactic acid which lowers the pH and in turn assists coagulation, promotes syneresis, helps prevent spoilage and pathogenic bacteria from growing, contributes

Fluid Milk Processing

to cheese texture, flavour and keeping quality. LAB also produce growth factors which encourages the growth of non-starter organisms, and provides lipases and proteases necessary for flavour development during curing. Further information on LAB and starter cultures can be found in the microbiology section.

After innoculation with the starter culture, the milk is held for 45 to 60 min at 25 to 30° C to ensure the bacteria are active, growing and have developed acidity. This stage is called ripening the milk and is done prior to renneting.

Milk Coagulation

Coagulation is essentially the formation of a gel by destabilizing the casein micelles causing them to aggregate and form a network which partially immobilizes the water and traps the fat globules in the newly formed matrix. This may be accomplished with:

- enzymes
- acid treatment
- heat-acid treatment.

Enzymes

Chymosin, or rennet, is most often used for enzyme coagulation.

Acid Treatment

Lowering the pH of the milk results in casein micelle destabilization or aggregation. Acid curd is more fragile than rennet curd due to the loss of calcium. Acid coagulation can be achieved naturally with the starter culture, or artificially with the addition of gluconodeltalactone. Acid coagulated fresh cheeses may include Cottage cheese, Quark, and Cream cheese.

Heat-Acid Treatment

Heat causes denaturation of the whey proteins. The denatured proteins then interact with the caseins. With the addition of acid, the caseins precipitate with the whey proteins. In rennet coagulation, only 76-78% of the protein is recovered, while in heat-acid coagulation, 90% of protein can be recovered. Examples of cheeses made by this method include Paneer, Ricotta and Queso Blanco.

Curd Treatment

After the milk has gel has been allowed to reach the desired firmness, it is carefully cut into small pieces with knife blades or wires. This shortens the distance and increases the available area for whey to be released. The curd pieces immediately begin to shrink and expel the greenish liquid called whey. This syneresis process is further driven by a cooking stage. The increase in temperature causes the protein matrix to shrink due to increased hydrophobic interactions, and also increases the rate of fermentation of lactose to lactic acid. The increased acidity also contributes to shrinkage of the curd particles. The final moisture content is dependant on the time and temperature of the cook stage. This is important to monitor carefully because the final moisture content of the curd determines the residual amount of fermentable lactose and thus the final pH of the cheese after curing.

When the curds have reached the desired moisture and acidity they are separated from the whey. The whey may be removed from the top or drained by gravity. The curd-whey mixture may also be placed in moulds for draining. Some cheese varieties, such as Colby, Gouda, and Brine Brick include a curd washing which increases the moisture content, reduces the lactose content and final acidity, decreases firmness, and increases openness of texture.

Curd handling from this point on is very specific for each cheese variety. Salting may be achieved through brine as with Gouda, surface salt as with Feta, or vat salt as with Cheddar. To achieve the characteritics of Cheddar, a cheddaring stage (curd manipulation), milling (cut into shreds), and pressing at high pressure are crucial.

Cheese Ripening

Except for fresh cheese, the curd is ripened, or matured, at various temperatures and times until the characteristic flavour, body and texture profile is achieved. During ripening, degradation of lactose, proteins and fat are carried out by ripening agents. The ripening agents in cheese are:

- bacteria and enzymes of the milk
- lactic culture
- rennet
- lipases
- added moulds or yeasts
- environmental contaminants.

Thus the microbiological content of the curd, the biochemical composition of the curd, as well as temperature and humidity affect

Yogurt

Yogurt (also spelled yogourt or yoghurt) is a semi-solid fermented milk product that originated centuries ago and has evolved from many traditional Eastern European (e.g., Turkish and Bulgarian) products. The word is from the Turkish *Yogen*, meaning *thick*. It's popularity has grown and is now consumed in most parts of the world.

Ingredients

Although milk of various animals has been used for yogurt production in various parts of the world, most of the industrialized yogurt production uses cow's milk. Whole milk, partially skimmed milk, skim milk or cream may be used. In order to ensure the development of the yogurt culture the following criteria for the raw milk must be met:

- low bacteria count
- free from antibiotics, sanitizing chemicals, mastitis milk, colostrum, and rancid milk
- no contamination by bacteriophages.

Other yogurt ingredients may include some or all of the following:

Other Dairy Products: concentrated skim milk, nonfat dry milk, whey, lactose. These products are often used to increase the nonfat solids content.

Sweeteners: glucose or sucrose, high-intensity sweeteners (e.g. aspartame).

Stabilizers: gelatin, carboxymethyl cellulose, locust bean Guar, alginates, carrageenans, whey protein concentrate.

Flavours

Fruit Preparations: including natural and artificial flavouring, colour.

Starter Culture

The starter culture for most yogurt production in North America is a symbiotic blend of *Streptococcus salivarius* subsp. *thermophilus* (ST) and *Lactobacillus delbrueckii* subsp. *bulgaricus* (LB). Although they can grow independantly, the rate of acid production is much higher when used together than either of the two organisms grown individually. ST grows faster and produces both acid and carbon dioxide. The formate and carbon dioxide produced stimulates LB growth. On the other hand, the proteolytic activity of LB produces stimulatory peptides and amino acids for use by ST. These microorganisms are ultimately responsible for the formation of typical yogurt flavour and texture. The yogurt mixture coagulates during fermentation due to the drop in pH. The streptococci are responsible for the initial pH drop of the yogurt mix to approximately 5.0. The lactobacilli are responsible for a further decrease to pH 4.0. The following fermentation products contibute to flavour:

- lactic acid
- acetaldehyde
- acetic acid
- diacetyl.

Manufacturing Method

The milk is clarified and separated into cream and skim milk, then standardized to achieve the desired fat content. The various ingredients are then blended together in a mix tank equipped with a powder funnel and an agitation system. The mixture is then pasteurized using a continuous plate heat exchanger for 30 min at 85° C or 10 min at 95° C. These heat treatments, which are much more severe than fluid milk pasteurization, are necessary to achieve the following:

- produce a relatively sterile and conducive environment for the starter culture
- denature and coagulate whey proteins to enhance the viscosity and texture.

The mix is then homogenized using high pressures of 2000-2500 psi. Besides thoroughly mixing the stabilizers and other ingredients, homogenization also prevents creaming and wheying off during incubation and storage. Stability, consistency and body are enhanced by homogenization. Once the homogenized mix has cooled to an optimum growth temperature, the yogurt starter culture is added.

A ratio of 1:1, ST to LB, inoculation is added to the jacketed fermentation tank. A temperature of 43° C is maintained for 4-6 h under quiescent (no agitation) conditions. This temperature is a compromise between the optimums for the two micoorganisms (ST 39°

C; LB 45° C). The titratable acidity is carefully monitored until the TA is 0.85 to 0.90%. At this time the jacket is replaced with cool water and agitation begins, both of which stop the fermentation. The coagulated product is cooled to 5-22° C, depending on the product. Fruit and flavour may be incorporated at this time, then packaged. The product is now cooled and stored at refrigeration temperatures (5° C) to slow down the physical, chemical and microbiological degradation.

Yogurt Products

There are two types of plain yogurt:
- Stirred style yogurt
- Set style yogurt.

The above description is essentially the manufacturing proceedures for stirred style. In set style, the yogurt is packaged immediately after inoculation with the starter and is incubated in the packages. Other yogurt products include:

Fruit-on-the-bottom style: fruit mixture is layered at the bottom followed by inoculated yogurt, incubation occurs in the sealed cups.

Soft-serve and Hard Pack Frozen Yogurt

Continental, French, and Swiss: stirred style yogurt with fruit preparation.

Yogurt Beverages

Drinking yogurt is essentially stirred yogurt which has a total solids content not exceeding 11% and which has undergone homogenization to further reduce the viscosity, Flavouring and colouring are invariably added. Heat treatment may be applied to extend the storage life. HTST pasteurization with aseptic processing will give a shelf life of several weeks at 2-4°C, which UHT processes with aseptic packaging will give a shelf life of several weeks at room temperature.

Other Fermented Milk Beverages

Cultured Buttermilk

This product was originally the fermented byproduct of butter manufacture, but today it is more common to produce cultured buttermilks from skim or whole milk. The culture most frequently used in S. lactis,, perhaps also spp. cremoris. Milk is usually heated to 95°C and cooled to 20-25°C before the addition of the starter culture. Starter is added at 1-2% and the fermentation is allowed to proceed for 16-20 hours, to an acidity of 0.9% lactic acid. This product is frequently used as an ingredient in the baking industry, in addition to being packaged for sale in the retail trade.

Acidophilus Milk

Acidophilus milk is a traditional milk fermented with Lactobacillus acidophilus (LA), which has been thought to have therapeutic benefits in the gastrointestinal tract. Skim or whole milk may be used.

The milk is heated to high temperature, e.g., 95°C for 1 hour, to reduce the microbial load and favour the slow growing LA culture. Milk is inoculated at a level of 2-5% and incubated at 37°C until coagulated. Some acidophilus milk has an acidity as high as 1% lactic acid, but for therapeutic purposes 0.6-0.7% is more common.

Another variation has been the introduction of a sweet acidophilus milk, one in which the LA culture has been added but there has been no incubation. It is thought that the culture will reach the GI tract where its therapeutic effects will be realized, but the milk has no fermented qualities, thus delivering the benefits without the high acidity and flavour, considered undesirable by some people.

Sour Cream

Cultured cream usually has a fat content between 12-30%, depending on the required properties. The starter is similar to that used for cultured buttermilk. The cream after standardization is usually heated to 75-80°C and is homogenized at >13 MPa to improve the texture. Inoculation and fermentation conditions are also similar to those for cultured buttermilk, but the fermentation is stopped at an acidity of 0.6%.

Others

There are a great many other fermented dairy products, including kefir, koumiss, beverages based on bulgaricus or bifidus strains, labneh, and a host of others. Many of these have developed in regional areas and, depending on the starter organisms used, have various flavours, textures, and components from the fermentation process, such as gas or ethanol.

Whipped Cream Structure

The structure of whipped cream is very similar to the fat and air structure that exists in ice cream. Cream is an emulsion with a fat content of 35-40%. When you whip a bowl of heavy cream, the agitation and the air bubbles that are added cause the fat globules to begin to partially coalesce in chains and clusters and adsorb to and spread around the air bubbles.

As the fat partially coalesces, it causes one fat-stabilized air bubble to be linked to the next, and so on. The whipped cream soon starts to become stiff and dry appearing and takes on a smooth texture. This results from the formation of this partially coalesced fat structure stabilizing the air bubbles. The water, lactose and proteins are trapped in the spaces around the fat-stabilized air bubbles. The crystalline fat content is essential (hence whipping of cream is very temperature dependent) so that the fat globules partially coalesce into a 3-dimensional structure rather than fully coalesce into larger and larger globules that are not capable of structure-building.

The structure of whipped cream as determined by scanning electron microscopy. A. Overview showing the relative size and prevalence of air bubbles (a) and fat globules (f); bar = 30 um. B. Internal structure of the air bubble, showing the layer of partially coalesced fat which has stabilized the bubble; bar = 5 um. C. Details of the partially coalesced fat layer, showing the interaction of the individual fat globules. Bar = 3 um.

Ice Cream

Ice cream has a long history as a popular dairy food item. It has evolved from a manually manufactured household product to a very automated industrial product. This is the ice cream homepage, a subset of the Dairy Technology Education Series. If you have come to this page directly, then you can go back to the beginning to start learning about dairy science and technology and dairy products.

Ice Cream: History and Folklore

Most of the following material has been extracted from "The History of Ice Cream", written by the International Association of Ice Cream Manufacturers (IAICM), Washington DC, 1978. As you will note below, however, much of the early history of ice cream remains unproven folklore.

Once upon a time, hundreds of years ago, Charles I of England hosted a sumptous state banquet for many of his friends and family. The meal, consisting of many delicacies of the day, had been simply superb but the "coup de grace" was yet to come.

After much preparation, the King's french chef had concocted an apparently new dish. It was cold and resembled fresh-fallen snow but was much creamier and sweeter than any other after-dinner dessert. The guests were delighted, as was Charles, who summoned the cook and asked him not to divulge the recipe for his frozen cream. The King wanted the delicacy to be served only at the Royal table and offered the cook 500 pounds a year to keep it that way. Sometime later, however, poor Charles fell into disfavour with his people and was beheaded in 1649. But by that time, the secret of the frozen cream remained a secret no more. The cook, named DeMirco, had not kept his promise.

This story is just one of many of the fascinating tales which surround the evolution of our country's most popular dessert, ice cream. It is likely that ice cream was not invented, but rather came to be over years of similar efforts. Indeed, the Roman Emperor Nero Claudius Caesar is said to have sent slaves to the mountains to bring snow and ice to cool and freeze the fruit drinks he was so fond of. Centuries later, the Italian Marco Polo returned from his famous journey to the Far East with a recipe for making water ices resembling modern day sherbets.

Most books are full of myths about the history of ice cream. According to popular accounts, Marco Polo (1254-1324) saw ice creams being made during his trip to China, and on his return, introduced them to Italy. The myth continues with the Italian chefs of the you Catherine de'Medici taking this magical dish to France when she went there in 1533 to marry the Duc d'Orleans, with Charles I rewarding his own ice-cream maker with a lifetime pension on condition that he did not divulge his secret recipe to anyone, thereby keeping ice cream as a royal perogative.

Unfortunately, there is no historical evidence to support any of these stories. They would appear to be purely the creation of imaginative nineteenth-century ice-cream makers and vendors. Indeed, we have found no mention of any of these stories before the nineteenth century.

They go on to refute the claims about Marco Polo, Catherine de'Medici, and Charles I (in particular, while the IAICM reference credits DeMirco as the Charles I chef, apparently while other various sources credit 10 different men, there are no records of such a pension being paid to any of Charles I's cooks).

They do go on in their book to discuss history for which there is a record, with (I think) the earliest written record being something made in China.

Chris Clarke, in his 2004 Royal Society of Chemistry mongraph "The Science of Ice Cream", points out quite correctly that the history of ice cream is closely associated with the development of refrigeration techniques and can thus be traced in several stages:

1. Cooling food and rink by mixing it with snow or ice;
2. The discosvery that dissolving salts in water produces cooling;
3. The discovery (and spread of knowledge) that mixing salts and snow or ice cools even further-mid to late 17th century-the inclusion of cream in the water ices also evolved around this time;
4. The invention of the ice cream maker in the mid 19th century;
5. The development of mechanical refrigeration in the later 19th and early 20th centuries-which led to the development of the modern ice cream industry.

Back to the IAICM History

In 1774, a caterer named Phillip Lenzi announced in a New York newspaper that he had just arrived from London and would be offering for sale various confections, including ice cream. Dolley Madison, wife of U.S. President James Madison, served ice cream at her husband's Inaugural Ball in 1813.

The first improvement in the manufacture of ice cream (from the handmade way in a large bowl) was given to us by a New Jersey woman, Nancy johnson, who in 1846 invented the hand-cranked freezer. This device is still familiar to many.

By turning the freezer handle, they agitated a container of ice cream mix in a bed of salt and ice until the mix was frozen. Because Nancy Johnson lacked the foresight to have her invention patented, her name does not appear on the patent records. A similar type of freezer was, however, patented on May 30, 1848, by a Mr. Young who at least had the courtesy to call it the "Johnson Patent Ice Cream Freezer".

Ice Cream Formulations

Ice Cream Mix General Composition:
- Milkfat: >10%-16%
- Milk solids-not-fat (snf): 9%-12%
- Sucrose: 10%-14%
- Corn syrup solids: 4%-5%
- Stabilizers: 0%-0.4%
- Emulsifiers: 0%-0.25%
- Water: 55%-64%.

The snf contains, on average, dry wt. basis, 38% protein, 54% lactose, and 8% ash (including 1.38% Ca, 1.07% P, 1.22% K, 0.7% Na).

Formulation Considerations

First of all, regulatory issues. What are the product definitions for your legal jurisdiction? These, of course, have to be met.

Next, desired Fat and Total solids (%):-Quality considerations, what kind of product are you trying to make?

Fat: MSNF Ratio's usually determined next based on fat content.

Sugar: Glucose solids Ratio's determined based on fat and total solids requirements, sweetness, freezing point depression, body and shelf life desired, and cost considerations.

Stabilizer/Emulsifier considerations come last, based on the ice cream formulation, and processing and distribution factors involved in each application.

With these considerations in mind, it is also useful to look at ranges of components for the categories of products in the *ice cream* category that are available in the market.

These are industry-used terms, not legally defined.

Economy Brands
- Fat content, usually legal minimum, e.g., 10%
- Total solids, usually legal minimum, e.g., 36%
- Overrun, usually legal maximum, ~120%
- Cost, low.

Standard Brands
- Fat content, 10-12%
- Total solids, 36-38%

Fluid Milk Processing

- Overrun, 100-120%
- Cost, average.

Premium Brands
- Fat content, 12-15%
- Total solids, 38-40%
- Overrun, 60-90%
- Cost, higher than average.

Super-premium Brands
- Fat content, 15-18%
- Total solids, >40%
- Overrun, 25-50%
- Cost, high.

Suggested Mixes

Suggested mixes for hard-frozen ice cream products.

	Percent (%)						
Milk Fat	10.0	11.0	12.0	13.0	14.0	15.0	16.0
Milk Solids-not-fat	11.0	11.0	10.5	10.5	10.0	10.0	9.5
Sucrose	10.0	10.0	12.0	14.0	14.0	15.0	15.0
Corn Syrup Solids	5.0	5.0	4.0	3.0	3.0	-	-
Stabilizer*	0.35	0.35	0.30	0.30	0.25	0.20	0.15
Emulsifier*	0.15	0.15	0.15	0.14	0.13	0.12	0.10
Total Solids	36.5	37.5	38.95	40.94	41.38	40.32	40.75

- Highly variable depending on type; manufacturers recommendations are usually followed.
- Usually an inverse relationship between fat and total solids compared to snf
- Generally an inverse relationship between glucose solids (corn sweetener) levels and total solids
- As total solids increases, there is less requirement for stabilizer
- As fat levels in a mix increase, there is generally less need for emulsifier.

Suggested mixes for low-fat (3-5% fat) and light (6-8% fat) ice cream products.

	Percent (%)				
Milk Fat	3.0	4.0	5.0	6.0	8.0
Milk SNF	13.0	12.5	12.5	12.0	11.5
Sucrose	11.0	11.0	11.0	13.0	12.0
CSS	6.0	5.5	5.5	4.0	4.0
Stabilizer	0.35	0.35	0.35	0.35	0.35
Emulsifier	0.10	0.10	0.10	0.15	0.15
Total Solids	33.65	33.45	34.45	35.5	36.0

Ice milk was the traditional lower fat ice cream product for many years, but this category has been re-classified by many regulatory jurisdictions to include three reduced fat categories: light ice cream, lowfat ice cream (the traditional ice milk), and non-fat ice cream. It has generally been possible to produce fat contents as low as 4% with traditional products, but further fat reductions have generally involved fat-replacers.

Table: Suggested mixes for soft-frozen ice cream products

	Percent (%)	
Milk Fat	10.0	10.0
Milk Solids-not-fat	12.5	12.0
Sucrose	13.0	10.0
Corn Syrup Solids	—-	4.0
Stabilizer*	0.35	0.15
Emulisifier*	0.15	0.15
Total Solids	36.0	36.3

*Highly variable depending on type; manufacturers recommendations are usually followed.

- Generally, while the fat content is kept lower, the snf content is generally higher than for hard-frozen products.
- Glucose solids are often used, but can lead to an enhanced sensation of guminess.
- Stabilizers are also generally used for viscosity enhancement and mouthfeel, but their function in ice recrystallization is no longer needed.

- Dryness, however, is a big concern in soft-serve products, hence the emulsifier content is generally kept high.

Sherbet and Sorbet

	Percent (%)	
Milk Fat	0.5	1.5
Milk Solids-not-fat	2.0	3.5
Sucrose	24.0	24.0
Corn Syrup Solids	9.0	6.0
Stabilizer/emulsifier*	0.3	0.3
Citric acid (50% sol.)**	0.7	0.7
Water	63.5	64.0
Total	100.0	100.0

- * Or as advised from the supplier.
- ** Acid is added just before freezing, after aging of the mix.
- Sorbet: delete the mix and skim powder
- Fruit: at about 25% to the mix.

Frozen Yogurt

	Percent (%)
Fat	2.0
MSNF	14.0
Sugar	15.0
Stabilizer	0.35
Water	68.65
Total	100.0

- Example Processing Instructions: 20% of this mix, consisting of skimmilk and skimmilk powder blended to give 12.5% solids, is to be incubated as the yogurt portion.
- To make the "incubated" portion, combine the appropriate amount of skimmilk and skimmilk powder, pasteurize at a high temperature, cool to 104 to 110F, and inoculate with a yogurt culture (typical of yogurt processing). When the fermentation is complete (to the desired acidity), cool the "yogurt".
- To make the "sweet" mix, combine the cream, sugar and stabilizer, and the balance of the skimmilk powder and skimmilk,

pasteurize, homogenize, cool (typical for ice cream processing), and blend with the "yogurt".
- The completed frozen yogurt mix is then aged and prepared for flavouring and freezing.

Ice Cream Ingredients

Ice cream has the following composition:
- greater than 10% milkfat by legal definition, and usually between 10% and as high as 16% fat in some premium ice creams
- 9 to 12% milk solids-not-fat: this component, also known as the serum solids, contains the proteins (caseins and whey proteins) and carbohydrates (lactose) found in milk
- 12 to 16% sweeteners: usually a combination of sucrose and glucose-based corn syrup sweeteners
- 0.2 to 0.5% stabilizers and emulsifiers
- 55% to 64% water which comes from the milk or other ingredients

These percentages are by weight, either in the mix or in the frozen ice cream. Please remember, however, that when frozen, about one half of the volume of ice cream is air, so by volume in ice cream, these numbers can be reduced by approximately one-half, depending on the actual air content.

However, since air does not contribute weight, we usually talk about the composition of ice cream on a weight basis, bearing in mind this important distinction. All ice cream flavours, with the possible exception of chocolate, are made from a basic white mix.

Formulations can be derived from a number of different starting points. Details and suggested formulas are detailed on the formulations page, but turning the formulation into a recipe depends on the ingredients used to supply the components, and it is then necessary to do a mix calculation to determine the required ingredients based on the formula. Ice milk and light ice creams are very similar to the composition of ice cream but in the case of ice milk in Canada, for example, it must contain between 3% and 5% milkfat by legal definition.

The ingredients to supply the desired components are chosen on the basis of availability, cost, and desired quality.

Milkfat (or "Butterfat") / Fat

Milkfat, or fat in general, including that from non-dairy sources, is important to ice cream for the following reasons:

- increases the richness of flavour in ice cream
- produces a characteristic smooth texture by lubricating the palate
- helps to give body to the ice cream, due to its role in fat destabilization
- aids in good melting properties, also due to its role in fat destabilization
- aids in lubricating the freezer barrel during manufacturing (Non-fat mixes are extremely hard on the freezing equipment).

The limitations of excessive use of butterfat in a mix include:

- cost
- hindered whipping ability
- decreased consumption due to excessive richness
- high caloric value.

The best source of butterfat in ice cream for high quality flavour and convenience is fresh sweet cream from fresh sweet milk. Other sources include butter or anhydrous milkfat.

During freezing of ice cream, the fat emulsion which exists in the mix will partially destabilize or churn as a result of the air incorporation, ice crystallization and high shear forces of the blades. This partial churning is necessary to set up the structure and texture in ice cream, which is very similar to the structure in whipped cream. Emulsifiers help to promote this destabilization process.

The triglycerides in milkfat have a wide melting range, +40° C to -40° C, and thus there is always a combination of liquid and crystalline fat. Alteration of this solid: liquid ratio can affect the amount of fat destabilization that occurs. Duplicating this structure with other sources of fat is difficult.

Vegetable (non-dairy) fats are used extensively as fat sources in ice cream in the United Kingdom, parts of Europe, the Far East, and Latin America but only to a very limited extent in North America. Five factors of great interest in selection of fat source are the crystal structure of the fat, the rate at which the fat crystallizes during dynamic temperature conditions, the temperature-dependent melting profile of the fat, especially at chilled and freezer temperatures, the content of high melting triglycerides (which can produce a waxy, greasy mouthfeel) and the flavour and purity of the oil.

It is important that the fat droplet contain an intermediate ratio of liquid:solid fat at the time of freezing. It is difficult to quantify this ratio as it is dependent on a number of composition and manufacturing factors, however, 1/2 to 2/3 crystalline fat at 4-5°C is a good, working rule. Crystallization of fat occurs in three steps: undercooling to induce nucleation, heterogeneous or homogeneous nucleation (or both), and crystal propagation.

In bulk fat, nucleation is predominantly heterogeneous, with crystals themselves acting as nucleating agents for further crystallization, and undercooling is usually minimal. However, in an emulsion, each droplet must crystallize independently of the next. For heterogeneous nucleation to predominate, there must be a nucleating agent available in every droplet, which is often not the case. Thus in emulsions, homogeneous nucleation and extensive undercooling may be common. Blends of oils are often used in ice cream manufacture, selected to take into account physical characteristics, flavour, availability, stability during storage and cost.

We have recently completed a study on the use of non-dairy fats in frozen desserts, which is available here. A blend of 75% of either fractionated palm kernel oil or coconut oil and 25% of an unsaturated oil, like high oleic sunflower oil, was shown to produce optimal levels of fat destabilization, meltdown and flavour, although coconut oil may take longer to crystallize during aging. Blends of 50% milkfat, 37.5% fractionated palm kernel or coconut oil, and 12.5% high oleic sunflower oil were also shown to be very acceptable.

Milk Solids-not-fat

The serum solids or milk solids-not-fat (MSNF) contain the lactose, caseins, whey proteins, minerals, and ash content of the product from which they were derived. They are an important ingredient for the following beneficialreasons:
- improve the texture of ice cream, due to the protein functionality
- help to give body and chew resistance to the finished product
- are capable of allowing a higher overrun without the characteristic snowy or flaky textures associated with high overrun, due also to the protein functionality
- may be a cheap source of total solids, especially whey powder

Fluid Milk Processing

The limitations on their use include off flavours which may arise from some of the products, and an excess of lactose which can lead to the defect of sandiness prevelant when the lactose crystallizes out of solution. Excessive concentrations of lactose in the serum phase may also lower the freezing point of the finished product to an unacceptable level.

The best sources of serum solids for high quality products are:
- concentrated skimmed milk
- spray process low heat skimmilk powder.

Other sources of serum solids include: sweetened condensed whole or skimmed milk, frozen condensed skimmed milk, buttermilk powder or condensed buttermilk, condensed whole milk, or dried or condensed whey. Superheated condensed skimmed milk, in which high viscosity is promoted, is sometimes used as a stabilizing agent but does, then, also contribute to serum solids.

It has recently become common practice to replace the use of skim milk powder or condensed skim with a variety of milk powder replacers, which are blends of whey protein concentrates, caseinates, and whey powders. These are formulated with less protein than skim powder, usually 20-25% protein, and thus less cost, but are blended with an appropriate balance of whey proteins and caseins to do an adequate job. Caution must be exercised in excessive use of these powders, experimentation with your own mix is the best answer.

See the section on Concentrated and Dried Dairy Products for a description of the manufacture of all of the above ingredients.

The proteins, which make up approximately 4% of the mix, contribute much to the development of structure in ice cream including:
- emulsification properties in the mix
- whipping properties in the ice cream
- water holding capacity leading to enhanced viscosity and reduced iciness.

Lactose Crystallization

1. A decrease in temperature favours rapid crystallization insofar as it increases the supersaturation.
2. A decrease in temperature favours slow crystallization insofar as it increases the viscosity, reduces the kinetic energy of the

particles, and decreases the rate of transformation from beta to alpha lactose.

Supersaturated state can exist, however, due to extreme viscosity, and it is likely that much of the lactose in ice cream is non-crystalline. Stabilizers help to hold lactose in supersaturated state due to viscosity enhancement. Fruits, nuts, candy-add crystal centres and may enhance lactose crystallization. Nuts pull out moisture from ice cream immediately surrounding the nut thus concentrating the mix. Citrate and phosphate ions decrease tendency for fat coalescence (Sodium citrate, Disodium Phosphate). They prevent churning in soft ice cream for example, producing a wetter product. These salts decrease the degree of protein aggregation. Calcium and magnesium ions have the opposite effect, promote partial coalescence. Calcium sulfate, for example, results in a drier ice cream. Calcium and Magnesium increase the degree of protein aggregation.

Salts may also influence electrostatic interactions. Fat globules carry a small net negative charge, these ions could increase or decrease that charge as they were attracted to or repelled from surface.

Sweeteners

A sweet ice cream is usually desired by the consumer. As a result, sweetening agents are added to ice cream mix at a rate of usually 12-16% by weight. Sweeteners improve the texture and palatability of the ice cream, enhance flavours, and are usually the cheapest source of total solids.

In addition, the sugars, including the lactose from the milk components, contribute to a *depressed freezing point* so that the ice cream has some unfrozen water associated with it at very low temperatures typical of their serving temperatures, -15° to -18° C. Without this unfrozen water, the ice cream would be too hard to scoop. The effect of sweeteners on freezing characteristics of ice cream mixes is demonstrated by the plot shown on the ice cream freezing curve.

Sucrose is the main sweetener used because it imparts excellent flavour. Sucrose is a disaccharide made up of glucose (dextrose, cerelose), and fructose (levulose). Sucrose is dextrorotatory-meaning it rotates a plane of polarized light to the right, + 66.5°. With hydrolyzed sucrose the plane of polarization is to the left, "inverted" -20°. An acid, plus water, plus heat treatment, at concentrations above 10%, yields invert sugar and increases the sweetness.

Fluid Milk Processing 153

It has become common in the industry to substitute all or a portion of the sucrose content with sweeteners derived from corn syrup. This sweetener is reported to contribute a firmer and more chewy body to the ice cream, is an economical source of solids, and improves the shelf life of the finished product.

Corn syrup in either its liquid or dry form is available in varying dextrose equivalents (DE). The DE is a measure of the reducing sugar content of the syrup calculated as dextrose and expressed as a percentage of the total dry weight. As the DE is increased by hydrolysis of the corn starch, the sweetness of the solids is increased and the average molecular weight is decreased. This results in an increase in the freezing point depression, in such foods as ice cream, by the sweetener. The lower DE corn syrup contains more dextrins which tie up more water in the mix thus supplying greater stabilizing effect against coarse texture.

An enzymatic hydrolysis and isomerization procedure can convert glucose to fructose, a sweeter carbohydrate, in corn syrups thus producing a blend (high fructose corn syrup, HFCS) which can be used to a much greater extent in sucrose replacement. However, these HFCS blends further reduce the freezing point producing a very soft ice cream at usual conditions of storage and dipping in the home.

Here is a diagram illustrating the effect of DE and maltose or fructose conversion on the properties of corn starch hydrolysates as used in ice cream.

A balance is involved between sweetness, total solids, and freezing point.

Stabilizers

The stabilizers are a group of compounds, usually polysaccharide food gums, that are responsible for adding viscosity to the mix and the unfrozen phase of the ice cream. This results in many functional benefits, listed below, and also extends the shelf life by limiting ice recrystallization during storage. Without the stabilizers, the ice cream would become coarse and icy very quickly due to the migration of free water and the growth of existing ice crystals.

The smaller the ice crystals in the ice cream, the less detectable they are to the tongue. Especially in the distribution channels of today's marketplace, the supermarkets, the trunks of cars, and so on, ice cream has many opportunities to warm up, partially melt some of the ice, and then refreeze as the temperature is once again lowered. This process is known as heat shock and every time it happens, the ice cream becomes more icy tasting. Stabilizers help to prevent this.

The functions of stabilizers in ice cream are:
- In the mix: To stabilize the emulsion to prevent creaming of fat and, in the case of carrageenan, to prevent serum separation due to incompatibility of the other polysaccharides with milk proteins, also to aid in suspension of liquid flavours
- In the ice cream at draw from the scraped surface freezer: To stabilize the air bubbles and to hold the flavourings, e.g., ripple sauces, in dispersion
- In the ice cream during storage: To prevent lactose crystal growth and retard or reduce ice crystal growth during storage, also to prevent shrinkage from collapse of the air bubbles and to prevent moisture migration into the package (in the case of paperboard) and sublimation from the surface
- In the ice cream at the time of consumption: To provide some body and mouthfeel without being gummy, and to promote good flavour release.

Limitations on their use include:
- production of undesirable melting characteristics, due to too high viscosity
- excessive mix viscosity prior to freezing
- contribution to a heavy or chewy body.

The stabilizers in use today include:

Locust Bean Gum : soluble fibre of plant material derived from the endosperm of beans of exotic trees grown mostly in Africa (Note: locust bean gum is a synonym for carob bean gum, the beans of which were used centuries ago for weighing precious metals, a system still in use today, the word carob and Karat having similar derivation).

Guar Gum: from the endosperm of the bean of the guar bush, a member of the legume family grown in India for centuries and now grown to a limited extent in Texas.

Carboxymethyl Cellulose (CMC): derived from the bulky components, or pulp cellulose, of plant material, and chemically derivatized to make it water soluble.

Xanthan Gum: produced in culture broth media by the microorganism *Xanthaomonas campestris* as an exopolysaccharide, used to a lesser extent.

Sodium Alginate: an extract of seaweed, brown kelp, also used to a lesser extent.

Carrageenan: an extract of Irish Moss or other red algae, originally harvested from the coast of Ireland, near the village of Carragheen but now most frequently obtained from Chile and the Phillipines.

Each of the stabilizers has its own characteristics and often, two or more of these stabilizers are used in combination to lend synergistic properties to each other and improve their overall effectiveness. Guar, for example, is more soluble than locust bean gum at cold temperatures, thus it finds more application in HTST pasteurization systems. Carrageenan is not used by itself but rather is used as a secondary colloid to prevent the wheying off of mix which is usually promoted by one of the other stabilizers.

Gelatin, a protein of animal origin, was used almost exclusively in the ice cream industry as a stabilizer but has gradually been replaced with polysaccharides of plant origin due to their increased effectiveness and reduced cost.

Emulsifiers

The emulsifiers are a group of compounds in ice cream that aid in developing the appropriate fat structure and air distribution necessary for the smooth eating and good meltdown characteristics desired in ice cream. Since each molecule of an emulsifier contains a hydrophilic portion and a hydrophobic portion, they reside at the interface between fat and water. As a result they act to reduce the interfacial tension or the force which exists between the two phases of the *emulsion*.

This causes a desorption of protein from the fat droplet surface, which promotes a destabilization of the fat emulsion (due to a weaker membrane) leading to a smooth, dry product with good meltdown properties. Their action will be more fully explained in the structure of ice cream section.

The original ice cream emulsifier was egg yolk, which was used in most of the original recipes. Today, two emulsifiers predominate most ice cream formulations:

Mono-and di-glycerides: derived from the partial hydrolysis of fats or oils of animal or vegetable origin.

Polysorbate 80: a sorbitan ester consisting of a glucose alcohol (sorbitol) molecule bound to a fatty acid, oleic acid, with oxyethylene groups added for further water solubility.

Other possible sources of emulsifiers include buttermilk, and glycerol esters. All of these compounds are either fats or carbohydrates, important components in most of the foods we eat and need. Together, the stabilizers and emulsifiers make up less than one half percent by weight of our ice cream. They are all compounds which have been exhaustively tested for safety and have received the "generally recognized as safe" or GRAS status.

Mix Calculations for Ice Cream and Frozen Dairy Desserts

The general objective in calculating ice cream mixes is to turn your formula into a recipe based on the ingredients you intend to use and the amount of mix you desire.

The formula is given as percentages of fat, milk solids-not-fat, sugar, corn syrup solids (glucose solids), stabilizers and emulsifiers. The ingredients to supply these components are chosen on the basis of availability, quality and cost.

The following table illustrates the relationship between the major components, the main ingredients that supply the major components, and the minor components that are supplied with the major ones for each ingredient.

Component and Ingredients to supply that component (but note that each of these ingredients also supplies the following other components): Milkfat, supplied by Cream (which also supplies SNF and water) or Butter (which also supplies SNF, water); Milk solids-not-fat (SNF, or sometimes also called serum solids, S.S.), supplied by any of the following:

- Skim powder (which also supplies water, about 3%)
- Condensed skim (which also supplies water)
- Condensed milk (which also supplies water and fat)
- Sweetened condensed (which also supplies water and sugar)
- Whey powder (which also supplies water).

Water, supplied by Skim milk (which also supplies msnf), or milk (which also supplies fat and msnf), or pure water.

Sweetener, supplied by dry or liquid (which also then supplies water) sucrose or corn syrup solids. The first step in a mix calculation is to identify for each ingredient we intend to use its components. If there is only one source of the component we need for the formula, for example the stabilizer or the sugar, we determine it directly by multiplying the percentage we need by the amount we need, e.g., 100 kg of mix @ 10% sugar would require 10 kg sugar. If there are two or more sources, for example we need 10 % fat and it is coming from both cream and milk, then we need to utilize an algebraic method.

Computer programs developed for mix calculations generally solve a simultaneous equation based on mass and component balances. To solve simultaneous equations, you need as many independent equations as you have unknowns.

For manual calculations, a method known as the "Serum Point" method has been derived. This method has solved the simultaneous equations in a general way so that only the equations need to be known and not resolved each time.

In standardizing mixes, the composition of the various ingredients used must be known. In some cases the percentage of solids contained in a product is taken as constant, while in others the composition must be obtained by analysis. Information on the various ingredients is given below:

(a) Skim milk-can be determined by analysis or assumed at 9 percent serum solids. Fat (0.01%-0.10%) should be taken into account if significant.

(b) Dried products, e.g. skim milk powder, whey powder, WPC, milk powder blends, usually taken to be 97 percent solids as they retain some moisture.

(c) Cream-Percent fat usually measured by an acceptable method. Percent MSNF found by formula as follows: (100-percent fat) x.09 = % snf (assuming that the "skim milk" contains 9% total solids). Example: In cream testing 30% fat, the percent snf would be (100-30) x.09 = 6.3% snf

(d) Milk-Percent fat measured by an acceptable method. Percent snf may be found same as for cream or by making a total solids test and deducting the percent fat.

(e) Condensed Milk Products-Composition of these products should be obtained by the supplier.

(f) Sweeteners-Sucrose-Dry 100% solids
Sucrose-Liquid 66% solids
Dextrose-Dry 100% solids
Corn Syrup Solids 100% solids
Corn Syrup Liquid 80% solids
Glucose 80% solids
Honey 80% solids

(g) Stabilizers and Emulsifiers (if solid)-Because of the small percentage used may be figured as 100 percent solids.

(h) Egg Products-Fresh whole eggs: 10% fat, 25% solids
Fresh egg yolk: 33% fat, 50% solids
Frozen egg yolk: 33% fat, 50% solids
Dried egg yolk: 60% fat, 100% solids

Below are some example problems to look at, if you are interested in the mathematics of mix calculations.

- Example 1. Basic mix using butter, skim powder, and water (only one source of each component). (Algebraic Method)
- Example 2. Mix using cream, skim, and skim powder (three sources of milk SNF, three sources of water). (Algebraic and Serum Point Methods)
- Example 3. Mix using cream, milk, and skim powder (three sources of milk SNF, three sources of water, and two source of fat). (Algebraic and Serum Point Methods)
- Example 4. Mix using cream, milk, and sweetened, condensed skim (three sources of milk SNF, three sources of water, two sources of fat, and two sources of sugar). (Serum Point Method)
- Example 5. Mix using cream, milk, and sweetened, condensed milk (three sources of milk SNF, three sources of water, three sources of fat, and two sources of sugar). (Serum Point Method)
- Example 6. Mix using a given amount of cream and skim, with the balance coming from butter, milk, and skim powder. (Serum Point Method)
- Example 7. Mix using cream, milk, condensed skim, and liquid sweeteners (water needs to be accounted for). (Serum Point Method)

After completing a problem, you should do a proof of your calculation, by ensuring that the mass sums to the desired value, and

Fluid Milk Processing

that the mass fraction of all components also sum to the desired value. There is only one unique solution, so you know by calculation if you have it right or not!

Ice Cream Manufacture

The basic steps in the manufacturing of ice cream are generally as follows:
- blending of the mix ingredients
- pasteurization
- homogenization
- aging the mix
- freezing
- packaging
- hardening.

Process flow diagram for ice cream manufacture: the red section represents the operations involving raw, unpasteurized mix, the pale blue section represents the operations involving pasteurized mix, and the dark blue section represents the operations involving frozen ice cream.

Blending

First the ingredients are selected based on the desired formulation and the calculation of the recipe from the formulation and the ingredients chosen, then the ingredients are weighed and blended together to produce what is known as the "ice cream mix". Blending requires rapid agitation to incorporate powders, and often high speed blenders are used.

Pasteurization

The mix is then pasteurized. Pasteurization is the biological control point in the system, designed for the destruction of pathogenic bacteria. In addition to this very important function, pasteurization also reduces the number of spoilage organisms such as psychrotrophs, and helps to hydrate some of the components (proteins, stabilizers).

Batch pasteurizers lead to more whey protein denaturation, which some people feel gives a better body to the ice cream. In a batch pasteurization system, blending of the proper ingredient amounts is done in large jacketed vats equipped with some means of heating, usually steam or hot water. The product is then heated in the vat to at least 69 C (155 F) and held for 30 minutes to satisfy legal requirements for pasteurization, necessary for the destruction of pathogenic bacteria.

Various time temperature combinations can be used. The heat treatment must be severe enough to ensure destruction of pathogens and to reduce the bacterial count to a maximum of 100,000 per gram. Following pasteurization, the mix is homogenized by means of high pressures and then is passed across some type of heat exchanger (plate or double or triple tube) for the purpose of cooling the mix to refrigerated temperatures (4 C). Batch tanks are usually operated in tandem so that one is holding while the other is being prepared. Automatic timers and valves ensure the proper holding time has been met.

Continuous pasteurization is usually performed in a high temperature short time (HTST) heat exchanger following blending of ingredients in a large, insulated feed tank. Some preheating, to 30 to 40 C, is necessary for solubilization of the components. The HTST system is equipped with a heating section, a cooling section, and a regeneration section. Cooling sections of ice cream mix HTST presses are usually larger than milk HTST presses. Due to the preheating of the mix, regeneration is lost and mix entering the cooling section is still quite warm.

Homogenization

The mix is also homogenized which forms the fat emulsion by breaking down or reducing the size of the fat globules found in milk or cream to less than 1 μ m. Two stage homogenization is usually preferred for ice cream mix. Clumping or clustering of the fat is reduced thereby producing a thinner, more rapidly whipped mix. Melt-down is also improved. Homogenization provides the following functions in ice cream manufacture:

- Reduces size of fat globules
- Increases surface area
- Forms membrane
- makes possible the use of butter, frozen cream, etc.

By helping to form the fat structure, it also has the following indirect effects:

- makes a smoother ice cream
- gives a greater apparent richness and palatability
- better air stability
- increases resistance to melting.

Homogenization of the mix should take place at the pasteurizing temperature. The high temperature produces more efficient breaking up of the fat globules at any given pressure and also reduces fat clumping and the tendency to thick, heavy bodied mixes. No one pressure can be recommended that will give satisfactory results under all conditions. The higher the fat and total solids in the mix, the lower the pressure should be. If a two stage homogenizer is used, a pressure of 2000-2500 psi on the first stage and 500-1000 psi on the second stage should be satisfactory under most conditions. Two stage homogenization is usually preferred for ice cream mix. Clumping or clustering of the fat is reduced thereby producing a thinner, more rapidly whipped mix. Melt-down is also improved.

Ageing

The mix is then aged for at least four hours and usually overnight. This allows time for the fat to cool down and crystallize, and for the proteins and polysaccharides to fully hydrate. Aging provides the following functions:

- Improves whipping qualities of mix and body and texture of ice cream.

It does so by:

- providing time for fat crystallization, so the fat can partially coalesce;
- allowing time for full protein and stabilizer hydration and a resulting slight viscosity increase;
- allowing time for membrane rearrangement and protein/emulsifier interaction, as emulsifiers displace proteins from the fat globule surface, which allows for a reduction in stabilization of the fat globules and enhanced partial coalescence.

Aging is performed in insulated or refrigerated storage tanks, silos, etc. Mix temperature should be maintained as low as possible without freezing, at or below 5 C. An aging time of overnight is likely to give best results under average plant conditions. A "green" or unaged mix is usually quickly detected at the freezer.

Freezing and Hardening

Following mix processing, the mix is drawn into a flavour tank where any liquid flavours, fruit purees, or colours are added. The mix then enters the dynamic freezing process which both freezes a portion of the water and whips air into the frozen mix. The "barrel" freezer is a scraped-surface, tubular heat exchanger, which is jacketed with a boiling refrigerant such as ammonia or freon. Mix is pumped through this freezer and is drawn off the other end in a matter of 30 seconds, (or 10 to 15 minutes in the case of batch freezers) with about 50% of its water frozen. There are rotating blades inside the barrel that keep the ice scraped off the surface of the freezer and also dashers inside the machine which help to whip the mix and incorporate air.

Ice cream contains a considerable quantity of air, up to half of its volume. This gives the product its characteristic lightness. Without air, ice cream would be similar to a frozen ice cube. The air content is termed its overrun, which can be calculated mathematically.

As the ice cream is drawn with about half of its water frozen, particulate matter such as fruits, nuts, candy, cookies, or whatever you like, is added to the semi-frozen slurry which has a consistency similar to soft-serve ice cream. In fact, almost the only thing which differentiates hard frozen ice cream from soft-serve, is the fact that soft serve is drawn into cones at this point in the process rather than into packages for subsequent hardening.

Hardening

After the particulates have been added, the ice cream is packaged and is placed into a blast freezer at -30° to -40° C where most of the remainder of the water is frozen. Below about -25° C, ice cream is stable for indefinite periods without danger of ice crystal growth; however, above this temperature, ice crystal growth is possible and the rate of crystal growth is dependant upon the temperature of storage. This limits the shelf life of the ice cream. A primer on the theoretical aspects of freezing will help you to fully understand the freezing and recrystallization process.

Hardening invloves static (still, quiescent) freezing of the packaged products in blast freezers. Freezing rate must still be rapid, so freezing techniques involve low temperature (-40oC) with either enhanced convection (freezing tunnels with forced air fans) or enhanced conduction (plate freezers).

Fluid Milk Processing

The rate of heat transfer in a frezing porcess is affected by the temperature difference, the surface area exposed and the heat transfer coefficient (Q=U A dT). Thus, the factors affecting hardening are those affecting this rate of heat transfer:

- Temperature of blast freezer-the colder the temperature, the faster the hardening, the smoother the product.
- Rapid circulation of air-increases convective heat transfer.
- Temperature of ice cream when placed in the hardening freezer-the colder the ice cream at draw, the faster the hardening;- must get through packaging operations fast.
- Size of container-exposure of maximum surface area to cold air, especially important to consider shrink wrapped bundles-they become a much larger mass to freeze. Bundling should be done after hardening.
- Composition of ice cream-related to freezing point depression and the temperature required to ensure a significantly high ice phase volume.
- Method of stacking containers or bundles to allow air circulation. Circulation should not be impeded-there should be no 'dead air' spaces (e.g., round vs. square packages).
- Care of evaporator-freedom from frost-acts as insulator.
- Package type, should not impede heat transfer-e.g., styrofoam liner or corrugated cardboard may protect against heat shock after hardening, but reduces heat transfer during freezing so not feasible.

Ice Cream Novelty/Impulse Products

Moulded Novelties

Ice cream novelties are typically single-serving items bought from vendors for immediate consumption. They come in a variety of shapes, sizes, colours, flavours, etc., and manufacturers are constantly developing new items to compete for market share. hence, this category is variably known as either impulse products or novelty products. These products can be manufactured by a number of processes. Products such as popsicles or ice cream with flat sides on sticks are manufactured by filling a mold with water ice or soft ice cream from a barrel freezer, immersion freezing that mold in a cold bath of calcim chloride solution, inserting the stick when partially hardened, and then after hardening is complete, removal of the ice cream from the mold by the stick.

Point 1 = filling;

Point 2 = partial freezing as the molds progress in the calcium chloride bath-you can have an optional "suck-out" of unfrozen material and refillng of something else at this stage;

Point 3 = stick insertion-the product is now frozen sufficiently to hold the stick in place;

Point 4 = further freezing as the molds progress in the calcium chloride bath-you can insert ribbon sauces anywhere along this stage, until the product gets too hard;

Point 5 = withdrawal of the mold from the calcium chloride bath, a quick defrost to free the bar from the mold wall, and extraction of the product from the mold by grabbing the stick-after which the product can be coated with nuts, enrobed with chocolate, then packaged, and on to storage. The molds continue underneath the machinewhere they are washed, rinsed, and sanitized, ready for the next filling.

Extruded Novelties

Products with no sticks, like bars with fancy shapes, or with sticks but irregular sides, such that they could not be pulled out of a mold, are frozen by extrusion. In such a process, the shape is formed and sliced, either vertically or horizontally as below, that slice is further hardenend by passing it through a cold tunnel, and then nuts or syrup can be coated on it, it can be enrobed with chocolate sauce, or whatever is desired to give the final product.

Horizontal extrusion is used to make the "chocolate bar analogue" type of products, while vertical extrusion is used to make everythng from chocolate-enrobed ice cream slices to fancy-shaped novelties on a stick.

Ice Cream Flavours

Most ice cream is purchased by the consumer on basis of flavour and ingredients. There are many different flavours of ice cream manufactured, and to some extent limited only by imagination. Vanilla accounts for 30% of the ice cream consumed. This is partly because it is used in so many products, like milkshakes, sundaes, banana splits, in addition to being consumed with pies, desserts, etc. It is the ice cream manufacturers responsibility to prepare an excellent mix,

but often they put the responsibility of the flavours and ingredients on the supplier. Ingredients are added to ice cream in four ways during the manufacturing process:
1. Mix Tank: for liquid flavours, colours, fruit purees, flavoured syrup bases Ð anything that will be homogeneously distributed in the frozen ice cream.
2. Variegating Pump: for ribbons, swirls, ripples, revels
3. Ingredient Feeder: for particulates-fruits, nuts, candy pieces, cookies, etc., some complex flavours may utilize 2 feeders
4. Shaker table: for large inclusions.

Generally, the delicate, mild flavours are easily blended and tend not to become objectionable at high concentrations, while harsh flavours are usually objectionable even in low concentrations. Therefore, delicate flavours are preferable to harsh flavours, but in any case a flavour should only be intense enough to be easily recognized. Flavouring materials may be:
1. Natural
2. Artificial or imitation
3. Blends of the two.

Table 1: US Ice Cream Consumption by Flavour, 2006

Type of Ice-cream	Percentage of Volume
1. Vanilla	30.2
2. Chocolate	10.0
3. Chocolate Chip	5.7
4. Butter Pecan	4.0
5. Strawberry	3.7
6. Neapolitan	3.0
7. Cookies and cream	2.6
8. Rocky Road	1.9
9. Cookie Dough	1.5
10. Cherry Vanilla	0.9
11. Coffee	0.7

Source: Dairy Facts, 2007, International Dairy Foods Association

Vanilla

Vanilla is without exception the most popular flavour for Ice Cream in North America. The dairy industry uses half of the total imported vanilla to North America. It is a very important ice cream ingredient, not only in vanilla ice cream, but in many other flavours where it is used as a flavour enhancer, e.g. chocolate much improved by presence of vanilla.

Vanilla comes from a plant belonging to the orchid family called Vanilla planifolia. There are several varieties of vanilla beans among which are Bourbon, Tahitian, Mexican. Bourbon beans are used to produce best vanilla extracts. Bourbons from Madagescar are the finest and account for over 60% of World production, Indonesia, 23%.

From each blossom of the vine that is successfully fertilized comes a pod which reaches 6-10 inches in length, picked at 6-9 months. It requires 26-29°C day and night throughout the season, and frequent rains with dry season near end for development of flavour.

Pods are immersed in hot water to "kill them" (also increases enzyme activity), then fermented for 3-6 months by repeated wrapping in straw to "sweat" and then uncovered to sun dry. 5-6 kg green pods produce 1 kg. cured pods. Beans then aged 1-2 yrs.

Enzymatic reactions produce many compounds-vanillin is the principal flavour compound. However, there is no free vanillin in the beans when they are harvested, it develops gradually during the curing period from glucosides, which break down during the fermentation and "sweating" of the beans. Extraction takes place as the beans are chopped (not ground) and placed in stainless steel percolator and warm alcohol (50°C, 50% solution) is pumped over and through the beans until all flavouring matter is extracted.

Concentrated Extract

Vacuum distillation takes place for a large part of the solvent. The desired concentration is specified as two fold, four fold, etc. Each multiple must be derived from an original 13.35 oz. beans.

Vanilla can be and is produced synthetically to a large extent. By-product of pulp and paper industry (lignan) or petrochemical industry (guaiacol). Compound flavours are produced from combination of vanilla extract and vanillin. Vanillin maybe added at one ounce to the fold and labelled Vanilla-Vanillin Flavour. Number of folds plus number oz. of vanillin equal total strength, eg. 2 fold + 2 oz. = 4 fold vanilla-vanillin. However, more than 1 oz to the fold is deamed imitation.

Vanilla flavouring is available in liquid form as:
- Natural Vanilla
- Natural and artificial (reinforced Vanilla with Vanillin)
- Artificial Vanilla (vanillin).

Usage level in the mix is a function of purity and concentration, usually ~0.3%.

Some vanillin actually improves flavour over pure vanilla extract but too much vanillin results in harsh flavours.

The choicest of ice creams can be made only with the best of flavouring materials. A good vanilla enhances the flavour of good dairy products in ice cream. It does not mask it.

Chocolate and Cocoa

The cacao bean is the fruit of the tree Theobroma cacao, (Cacao, food of the gods) which grows in tropical regions such as Mexico, Central America, South America, West Indies, African West Coast. The word cocoa is a corruption of the native word cacao. The beans are embedded in pods on the tree, 20-30 beans per pod. When ripe, the pods are cut from the trees, and after drying, the beans are removed from the pods and allowed to ferment, 10 days (microbiological and enzymatic fermentation). Beans then are washed, dried, sorted, graded and shipped.

At the processing plant, beans are roasted, seed coat removed-called the nib. The nib is ground, friction melts the fat and the nibs flow from the grinding as a liquid, known as chocolate liquor.

Liquor: 55% fat, 17% carbohydrate, 11% protein, 6% tannins and many other compounds (bitter chocolate-baking).

Cocoa butter: fat removed from chocolate liquor, narrow melting range 30 to 36° C.

Cocoa: after the cocoa butter is pressed from the chocolate liquor, the remaining press cake is now material for cocoa manufacture.

The amount of fat remaining determines the cocoa grade:
- medium fat (Breakfast) cocoa 20-24% fat
- low fat 10-12% fat.

Cocoa powder can also be alkalized, which reduces acidity/astringency and darkens the colour. Slightly alkalized cocoa is usually preferred in ice cream because it gives a deeper colour but the choice depends upon:

- consumer preference
- desired colour (Blackshire cocoa may be used to darken colour)
- strength of flavour
- fat content.

There are many types of chocolate that differ in the amounts of chocolate liquor, cocoa butter, sugar, milk, other ingredients, and vanilla.

Imitation Chocolate

Replacing some or all of the cocoa fat with other vegetable fats. Improved coating properties, resistance to melting.

White Chocolate

Cocoa butter, MSNF, sugar, no cocoa or liquor.

In chocolate ice cream manufacture, cocoa is more concentrated for flavouring than chocolate liquor (55% fat) because cocoa butter has relatively low flavour. However, the cocoa fat adds texture to the ice cream. Acceptable mixes can be made using 3% cocoa powder, 2.5% cocoa powder plus 1.5% chocolate liquor, or 5% chocolate liquor.

A good chocolate ice cream will be made if the cocoa and/or chocolate liquor is added to the vat and homogenized with the rest of the mix. Chocolate mixes have a tendency to become excessively viscous so stabilizer content and homogenizing pressure need to be adjusted.

One problem is called chocolate specking. It can occur in soft serve ice cream, when cocoa fibres become entrapped in the churned fat.

Fruit Ice Cream

Fruit for Ice Cream is available in the following forms:
1. Fresh Fruit
2. Raw Frozen Fruit
3. Open Kettle Processed Fruit
4. Aseptically Processed Fruit.

Advantages of processed fruits:
1. Purchasing year round supply: problems of procurement and storage transferred to fruit processor

Fluid Milk Processing

2. Availability: blending of sources from around the world in RTU form, no thawing, straining, etc.
3. Quality control: processor adjusts for quality variations
4. Ice Cream quality: fruit won't freeze in ice cream, usually free of debris, straw, pits.
5. Microbial Safety
6. Convenience.

Fruit feeders are used with continuous freezers to add the fruit pieces, while any fruit juice is added directly to the mix. Fruit is usually added at about 15-25% by weight.

Nuts in Ice Cream

Nuts are usually added at about 10% by wt. Commonly used are walnuts, pecans, filberts, almonds and pistachios. Brazil nuts and cashews have been tried without much success.

Quality Control of Nutmeats for Ice Cream

1. Extraneous and Foreign Material: Requires extensive cleaning, Colour Sorter, Destoner, X-rays, Aerator, Hand-Picking, Screening
2. Microbiological Testing: Aflatoxin contamination can be a hazard with Peanuts, Pistachios, Brazils. All nutmeats should receive random testing for: Standard Plate Count, Coliform, E. Coli, Yeast and Mold, Salmonella.
3. Bacteria Control: Nuts must be processed in a clean sanitary premise following good manufacturing practices. Nuts should be either oil roasted or heat treated to reduce any bacteria.
4. Sizing: Some nutmeats require chopping to achieve a uniform size in order to fit through the fruit feeder, i.e.: Pecans, Almonds, Peanuts, Filberts
5. Storage Nutmeats should be stored at 34-38° F to maintain freshness and reduce problems with rancidity.

Colour in Ice Cream

Ice cream should have a delicate, attractive colour that suggests or is closely associated with its flavour. Almost all ice creams are slightly coloured to give them the shade of the natural product 15% fruit produces only a slight effect on colour. However, most suppliers, would include some colour in the fruit to save the processor time i.e. solid pack strawberries include colour. Most colours are of synthetic origin, must be approved, purchased in liquid or dry form. Solutions can easily become contaminated and therefore must be fresh.

Colours are used in ice cream to create appeal. If used to excess they indicate cheapness. The choice of shade is dictated by flavour, i.e. red for strawberry, light green for mint, purple for grape, etc.

Homemade Ice Cream

Ice cream fills a useful place in homes throughout the country. It is a favourite for desserts or snacks incorporating an array of many flavour variations. Ice cream contains many nutrients. With the recipes provided, all should be able to enjoy some type of this tempting food. If the regular recipe does not suit your needs, there is the low calorie recipe which contains less than 3% fat for both a cost and calorie saving.

The recipe using coffee whitener is significantly less costly than the regular and does not contain milk fat should that be your limitation. So let's mix up a batch of ice cream for anyone and everyone to enjoy!

Ingredients Used

The main constituents of ice cream are fat, milk solids-not-fat (skim-milk powder), sugar, gelatin (or other suitable stabilizer), egg and flavouring.

A variety of milk products can be used: cream, whole milk, condensed milk and instant skim-milk powder. The recipes stated below proved satisfactory using whipping cream (32-35% fat), table cream (18% fat) and whole milk. The fat gives the product richness, smoothness and flavour. Skim-milk powder is used to increase the solids content of the ice cream and give it more body. It is also an important source of protein which will improve the ice cream nutritionally. Use good quality, fresh powder to avoid imparting a stale flavour to the ice cream.

Liquid coffee whitener (usually purchased frozen) is a cream substitute in one of the recipes. It will yield a slightly different flavour which is still very acceptable. The texture of the ice cream is very creamy. Liquid coffee whitener offers the convenience of being stored frozen in your freezer and is readily available if a quick decision is made to make ice cream.

Fluid Milk Processing 171

Sugar is a common ingredient to use as a sweetener. It increases the palatability and improves the body and texture.

The next ingredient, gelatin (or similar substance) assists in absorbing some of the free water in the ice cream mix and helps prevent the formation of large crystals in the ice cream.

It also gives substance or a less watery taste when the ice cream is consumed. The eggs are added to make the fat and water more miscible and also to improve the whipping ability which gives the ice cream greater resistance to melting.

Preparation of the Ice Cream Mix

The mix (unfrozen ice cream) has to be cooked (pasteurized). For pasteurizing the mix, it is best to use a double boiler to prevent scorching.

Place the liquid ingredients (milk, cream or coffee whitener) in the upper section of the double boiler. Beat in the eggs and the skim-milk powder. Mix the gelatin with the sugar and add to the liquid with constant mixing. While stirring, heat to about 70°C. Place the container in cold water and cool as rapidly as possible to below 18°C.

Aging the Mix

The ice cream mix is best if it is aged (stored in the refrigerator) overnight. This improves the whipping qualities of the mix and the body and texture of the ice cream. If time does not permit overnight aging, let the mix stand in the refrigerator for at least four hours. After the aging process is completed, remove the mix from the refrigerator and stir in the flavouring.

Freezing the Mix

The freezing procedure has a two-fold purpose, the removal of heat from the mix and the incorporation of air into the mix. Heat is removed by conduction through the metal to the salt water brine surrounding the freezing can. This transfer of heat depends upon the temperature of the brine, the speed of the dasher and how well the dasher scrapes the cold mix from the surface of the freezer can. The dasher speed and surface contact are important to achieve complete removal of the frozen ice cream from the wall of the freezer can. A brine made from 500 grams (1 lb.) of salt and 5 kilograms (11 lbs.) of crushed ice (one pail full) makes a good freezing mixture.

Before starting to freeze the ice cream, make sure all parts of the freezer coming in contact with the ice cream are clean and have been scalded. Let the can cool before pouring in the mix. Place the empty can in the freezer bucket and insert the dasher ensuring both the can and the dasher are centred. Pour the cold, aged mix into the freezer can. The can should not be filled over two-thirds full to allow sufficient room for air incorporation.

The recipes listed below will fill a 5 litre (5 quart U.S.) freezer can to just below the fill line. Attach the motor or crank mechanism, depending on whether your freezer is the electric or hand-cranked style, and latch down securely. Plug in the motor or start turning the crank. Immediately begin adding crushed ice around the can sprinkling it generously with salt. Try to add the salt and ice in the same one to ten proportion to get the proper brine temperature. After the bucket is filled with ice to the overflow hole, pour a little water over the ice to aid in the melting process.

Freeze the mix for 20 to 30 minutes. If the electric motor stalls, immediately unplug it. Remove the motor or crank and take the dasher out of the ice cream. The ice cream will be softly frozen. Scrape the ice cream from the dasher and either scoop into suitable containers or pack in the freezer can. Immediately place the ice cream in the deep freeze to harden.

If freezer facilities are not available, the ice cream can be left in the can, the lid plugged with a cork and placed back into the bucket. Repack the freezer with more ice and salt, cover with a heavy towel and set in a cool place to harden until serving time. This will require further addition of ice and salt depending on the length of time the ice cream is being held. The yield from the recipes listed below should be three to four litres.

Regular Vanilla Ice Cream

Table cream 2 litres (2 US quarts)

Instant skim-milk powder 350 ml (1.5 cups)

Sugar 450 ml (2 cups)

Gelatin one 7 g (1/4 oz.) pkg.

Egg one med or large

Vanilla 10 ml (2 teaspons)

Calories per 100 g 230

Low Calorie Vanilla Ice Cream

Whole milk 2 litres (2 US quarts)

Instant skim-milk powder 500 ml (2 cups)

Sugar 350 ml (1.5 cups)

Gelatin one 7 g (1/4 oz.) pkg.

Egg one med or large

Vanilla 10 ml (2 teaspons)

Calories per 100 g 125

Milk Substitute Vanilla Ice Cream

Coffee whitener 2 litres (2 US quarts)

Instant skim-milk powder 350 ml (1.5 cups)

Sugar 500 ml (2 cups)

Gelatin one 7 g (1/4 oz.) pkg.

Egg one med or large

Vanilla 10 ml (2 teaspons)

Calories per 100 g 210

Hints for Making Good Ice Cream

1. If the ice cream is very soft, the brine is not cold enough. More salt should be added to reduce the brine temperature.
2. If the ice cream is coarse and ice in less than 20 minutes, the brine has become to too cold too quickly. Too much salt has been used.
3. Make the ice cream mix the day before it is frozen to get a smoother product and a higher yield.
4. Electric freezing takes longer than hand operated.
5. Use crushed ice for freezing.
6. Freeze at least 3 hours before the ice cream is to be served.
7. Be sure dasher is properly centred in the freezer can.
8. Add liquid flavours before freezing but if you want to add fruits or nuts, add them after freezing and before hardening.
9. Use a wire whip to blend ingredients for best results.
10. Clean the salt off all the metal parts of the freezer to prevent corrosion.

Milk and Cheese

In food preparation, milk is used in many ways and combined with many kinds of foods. All the different types of cheese are made from it. Meat, vegetables, and cereals may be cooked in it. It is used as a basis for many sauces. Such sauces may be combined with eggs, with meats, or with vegetables; it is used in puddings and in frozen desserts; it is used for soups and for drinks like cocoa and coffee; it is combined with cereals; it is used in combination with many foods, as in custards, in cakes, and in quick breads.

Milk from different animals is used for food, but in this country unless the source of the milk is mentioned it is understood to be cow's milk.

Composition Of Milk

Milk from different animals varies somewhat in the proportion of the different constituents. All milk contains a high proportion of water, cow's milk averaging about 87 per cent. The milk from different breeds of cattle varies considerably in composition. The milk of individual cows of the same breed also varies in the proportion of the different constituents. This may be partly due to environment, partly to inheritance, and partly to individuality. Milk may vary particularly in fat content from one milking to the next. The percentage of fat increases during the milking period; that is, the first milk obtained is not so rich in fat as the last portions of milk.

The fat varies for different breeds from about 3.5 per cent for Holstein to about 5 per cent for Guernsey and Jersey.

The protein varies from about 3.3 per cent for Holstein to 4.0 per cent for Guernsey and Jersey. It parallels the fat content somewhat, being highest in the breeds having a high fat content and lowest in those having a low fat content.

The lactose does not vary so much with the different breeds as the fat and protein. The lactose content is from about 4.65 to 5 per cent.

The ash varies from about 0.68 to 0.75 for the different breeds.

Chemical And Physical Properties

The constituents of milk that are most important in food preparation are enzymes, vitamins, pigments, salts, sugar, fat, and proteins.

Enzymes: The enzymes of cow's milk are reported as follows by Rogers; proteinases, lactase, diastase, lipase, salolase, catalase, peroxidase, and aldehydrase. Rogers states that the proteolytic enzyme, galactase, brings about slow decomposition of milk proteins into peptones, amino acids, and ammonia.

Vitamins: All the vitamins recognized at the present time are contained in milk. Some are present in comparatively large and others in smaller amounts.

Pigments: The appearance of milk is white. This is due to light rays reflected by the colloidally dispersed constituents of the milk, the calcium caseinate, and calcium phosphate.

Milk contains two classes of yellow or orange pigments. The water-soluble pigment, which imparts a yellow colour with a green fluorescence to the whey of milk, was formerly called lactochrome. A name recently suggested for this pigment is lactoflavin. It is regarded as one flavin of a specific group, collectively to be called lyochromes. It is possible that lacto-flavin is composed of more than one pigment. Rogers says "lactoflavin forms compounds with saccharides, proteins, and purines (uric acid). These compounds possibly either occur naturally in milk or readily form during isolations, thus accounting for the several lactoflavins isolated from milk." It is probable that the pigment lactoflavin is one of the five fractions of vitamin G (B2). Milk is relatively rich in this vitamin.

A fat-soluble pigment, carotene, found in the fat gives the milk a more or less yellow tinge, which is more pronounced as the fat particles become more concentrated and form cream. The group of pigments called caroti-noids, which include carotene, xanthophyll, and related pigments. The chief pigment of butter fat is the carotene, but little xanthophyll being found. The depth of colour depends upon the amount of pigment present. The colour of carotene in solution varies from yellow to orange and to a deep red-orange as the concentration increases.

The amount of carotene found in the butter fat depends upon the extent of carotene in the food of the cow. Green grasses, hay cured to retain its green colour, green corn, and carrots are rich in carotene. The carotene content of milk fat is less rich during the winter months, if the food of the cow is poor in carotene during this period. Only in cow's milk is the fat extensively pigmented. With the exception of the fat of human milk, which is often pigmented, the fat of the milk of other animals is either devoid of or contains little pigment.

Salts: Milk contains salts of potassium, sodium, magnesium, calcium, phosphates, chlorides, and citrates. Traces of sulfates and carbonates are found. Iron is present in small amount. Iodides are also found in small amounts, the amount being greater in some localities than in others. Iodides may be easily transmitted from the feed to the milk. Supplee and Bellis have found copper to average about 0.52 part per million in freshly drawn milk. Brickner has reported that milk contains 3.6 to 5.6 parts of zinc per million parts of milk. Manganese in normal milk averages 0.02 to 0.06 parts per million. The greater part of the sulfur is found in the milk proteins. Barger and Coyne state that part of the sulfur in milk is found in the amino acids methionine and cystine of the proteins, but that all of the sulfur is not accounted for.

The salts of milk are found in milk in solution, in the colloidal state, and in combination with the proteins. The exact chemical combinations of the different salts in the milk are not fully determined. For this reason different authorities report different salt combinations. Thus formulas imply definite combination, whereas there is a complex salt equilibrium in milk, which has not been satisfactorily worked out. The citrates, the combinations of which may be trisodium and tripotassium citrate, tricalcium and trimagnesium citrate, are probably entirely in solution. The possible chlorides of potassium, sodium, and calcium are also in solution. Some authorities believe that the phosphates are present chiefly as dicalcium phosphate, $CaHPO_4$ others believe that tricalcium phosphate, $Ca_3(PO_4)_2$, is the principal phosphate combination. The phosphates are partly in solution, with the greater portion in colloidal dispersion. When the particles of dicalcium phosphate are heated they become aggregated and partially precipitated.

Calcium and magnesium are in combination with the casein to form calcium and magnesium caseinates. Zoller states that there may be traces of sodium and potassium caseinates.

Lactose: The solubility of lactose and its properties may be found in the chapter on sugar. It is caramelized by heat at rather low temperatures.

Fat: Butter fat is composed of glycerol and fatty acids. Fatty acids of both the saturated and unsaturated series are present. The relative percentage of the unsaturated fatty acids varies with the feed, averaging higher in summer than in winter. Dean and Hilditch state that the oleic and linoleic acids increase by 4 per cent (mols), with a parallel diminution in butyric and stearic acids when cows return to pasture. They also report that with increased age of the cow the unsaturated acids increase at the expense of palmitic acid, which was lowered from 29 to 22-23 per cent by weight of the total fat in the four years of observations made on milk from the same group of cows. The saturated fatty acids are butyric, caproic, caprylic, capric, lauric, myristic, plamitic, and stearic. Arachadonic acid has been reported absent by some investigators. Butter fat contains a higher proportion of the first-named saturated acids than other food fats. The first ones in the series are quite volatile with steam, the volatility decreasing with increase in molecular weight of the acid. Hence, when butter is heated for several minutes, the percentage of the lower saturated fatty acids may be decreased. The unsaturated fatty acids include oleic, linoleic, and arachidic.

Formerly it was thought that the particular flavour of butter was due to the greater proportion of the lower saturated fatty acids, but now it is known that butters with satisfactory flavour and aroma (Michaelian, Farmer, and Hammer) contain considerable quantities of acetylmethyl-carbinol plus diacetyl. The acetylmethylcarbinol in a pure condition is odorless, but by bacterial action it is changed to diacetyl, which in concentrations of 0.0002 to 0.0004 per cent in or added to neutral butter gives a characteristic aroma.

Size of fat Globules: The fat in milk is found in small globules. These fat globules which are microscopic in size are suspended in the milk. They vary in size from 0.10μ to 22.0μ. Most of the globules are less than 10μ and average about 3μ in diameter. The size of the fat globules varies with different breeds, being larger in milk from Jersey and Guernsey than in milk from other breeds, (2) with the lactation period, decreasing in size with length of the lactation period, and (3) with feed. Dry feed tends to decrease the size of the globules, succulent feed to increase their size.

Creaming: The specific gravity of the fat globules is less than that of the fluid of the milk. Hence there is a tendency for them to rise to the surface to form a cream layer. The extent of creaming depends upon several factors, such as the size of the fat globules, temperature of the milk, acidity, physical state of the fat, etc. Creaming occurs more rapidly in milk when the fat globules are quite large than when they are smaller. In rising to the top of the milk the globules of fat clump together, and this clumping increases the tendency for them to rise to the surface of the milk. As the larger clumps rise they may carry many of the smaller globules to the surface. Therefore, clumping not only aids the completeness of creaming but also the rate, for the rate is more rapid when the fat particles clump quickly. Rogers states that the factors affecting clumping are "temperature, the acidity, the fat content and its degree of dispersion, the degree of agitation, and the fluidity of the system. The fat content, the degree of agitation and the fluidity of the system determine the probability of collisions of the globules." The tendency of the fat globules to clump is greatest when the milk is cooled rapidly to 7° to 8°C, but if the fat globules become solid at the low temperature before creaming is allowed to take place the rate of creaming is retarded. The temperature that favours clumping is also best for whipping the cream. Most cream is separated from the milk by mechanical means. The fat particles left in the milk after separating the cream are those less than 1u and those between 1 and 2u in diameter.

Butter: The fat globules in cow's milk are suspended in the milk and thus do not form a permanent emulsion, though they may be so reduced in size by homogenization that they form a permanent emulsion. Of the different theories formulated for explaining the manner in which emulsions are stabilized the adsorption film theory is usually connected with milk. Substances that lower surface tension tend to collect at the interface between two non-miscible systems. Proteins tend to lower the surface tension, hence tend to collect at the interface. The layer or film around the fat globules probably consists of adsorbed calcium caseinate, with some lactal-bumin, globulin, and calcium phosphate. This membrane may be weakened or broken in various ways. When milk is heated slowly, the membrane surrounding some of the fat globules may be broken and a number of globules may coalesce.

Sometimes milk that has been heated and then cooled has a more oily appearance due to this coalescing of the fat globules. The membrane surrounding the fat particles may be broken by mechanical agitation. Formation of butter in churning is brought about in this manner.

There are two theories regarding butter formation. One is that the emulsion is reversed from the type found in the milk and that the butter is a water-in-fat emulsion.

The other view is that butter is formed by packing the fat globules into a compact mass, and that water and air are enmeshed during this process. Temperature and the formation of a foam are both important in churning. At a temperature above 65°C. there is no aggregation of fat particles. Below 65° the tendency to clump increases and is at a maximum at 7° to 8°C. A favourable temperature for butter formation is below 24° and above 10°C. At temperatures below 4°C. the fat globules do not adhere to each other and aggregation does not take place. At temperatures above the melting point of butter, no butter is formed.

The fat particles tend to clump at the liquid/air interface, so that air beaten in during the churning process accelerates clumping of the fat. Butter may be churned from sweet or sour cream. The flavour of butter from the sweet cream is milder and different from that of the sour-cream butter. Many housekeepers prefer the sweet-cream butter for table use and the sour-cream butter for cooking.

The adsorbed films surrounding the fat globules may also be destroyed by the addition of acid or alkali. It is by these methods that the fat is set free for a quantitative determination. In the Babcock test, acid is used for liberating the fat globules; in the Hoyberg test alkali is used. The Bab-cock, or some modification of it, is the one usually employed in estimating the fat content of milk and cream.

Protein: The chief proteins found in milk in order of their decreasing amounts are casein, lactalbumin, and lactoglobulin.

Casein: Casein belongs to the group of phosphoproteins. The form in which the phosphorus exists in the casein is not definitely known, but it is believed to be present in the form of combined phosphoric acid. Casein forms about 3 per cent of cow's milk.

At its isoelectric point, which is pH 4.6, casein is nearly insoluble in water. Casein is amphoteric and forms salts with acids and alkalies. Fresh milk has a reaction of about pH 6.6, so that the casein is present in the milk as salts of bases and is found as calcium and magnesium caseinates.

All the alkali caseinates are soluble in water, though the salts of the alkaline earths are less soluble than the alkali ones. Loeb states that below pH 4.6 the casein chloride, casein acetate, and casein lactate are very soluble in water, but casein sulfate and casein oxalate are difficultly soluble.

According to Zoller, pure casein when heated in water begins to imbibe water at 80° to 90°C. and becomes plastic. In this form it can be moulded and shaped. Upon cooling it becomes very hard. Casein can be precipitated from milk by bringing the milk to the isoelectric point of casein. Coagulation of casein will be considered later.

Lactalbumin: The proportion of lactalbumin in milk is much lower than that of casein. It forms about 0.50 per cent of cow's milk. Its isoelectric point is pH 4.55. Since the reaction of fresh milk is about pH 6.6, it is on the alkaline side of the isoelectric point of lactalbumin. Thus is it possible that the lactalbumin is found combined as salts of bases, such as calcium and magnesium albuminates. Osborne and Wakeman think it is uncombined. Lactalbumin is soluble in water, and is coagulated by heating in solution to a temperature of about 70°C. Coagulation may not be complete at this temperature. Palmer states that the lactalbumin is more highly dispersed than the other colloidal constituents of the milk.

Lactoglobulin: Lactoglobulin occurs in milk in very small quantities, about 0.05 per cent of cow's milk. Lactoglobulin is insoluble in distilled water, but it is soluble in dilute solutions of strong bases or acids, and in dilute salt solutions. It is coagulable by heat.

Homogenization of Milk

Globules of butter fat are suspended in the milk. They are surrounded by films of adsorbed caseinates, albuminates, and globulinates. The fat globules of milk are too large to form a permanent emulsion, so they gradually rise to the top of the milk in the form of cream. If the milk or cream is put through a machine called a homogenizer, the fat globules are reduced in size. This is accomplished by using pressure and forcing the milk or cream through small openings. Homogenized milk or cream may form a stable emulsion if the fat globules are reduced enough in size. Hence, when the fat is broken into fine enough globules the cream will not rise to the top of the homogenized milk.

The size of the fat globules after homogenization depends upon the temperature of the milk during homogenization and the pressure used. With increase in temperature the degree of dispersion increases rapidly from 40° to 65°C, so that the smallest fat particles are obtained at 65°. Ordinarily temperatures above 65° are not used for homogenization. The size of the fat particles also decreases with increased pressure.

Whipped cream is stabilized by proteins. The fat globules in cream are surrounded by films of protein substances. Homogenized cream also has the film of adsorbed proteins around the fat particles. Clayton states the fat particles in homogenized cream may be 1000 times greater in number than before homogenization. Since the number of fat particles is increased, the amount of globulinates, caseinates, and albuminates used in forming films is very much greater, for the surface area of the fat globules has increased enormously.

Whipped cream is both an emulsion and a foam. The fat particles must be surrounded by a film of protein in order to be stabilized, and the air globules must be surrounded by a film of protein to stabilize them. In homogenized cream most of the protein is used in surrounding the fat globules, on account of the increased surface area of the smaller and increased number of fat globules, and thus there is not enough left to surround the air bubbles. Hence, homogenized cream seldom whips unless protein is added for film forming.

Factors that affect the whipping quality of cream. In addition to the protein or film forming in whipping cream, the fat content, the size of the fat particles, the temperature of whipping, and the viscosity are important factors. Dahlberg and Hening have studied the relation of viscosity, surface tension, and whipping properties of milk and cream. They have found that increased viscosity increases the whipping properties of cream, but the lowering of the surface tension does not improve the whipping qualities. They have reported two changes taking place during whipping. The incorporation of air depends upon the milk proteins forming the film around the air globules, and the rigidity or stiffness of the whipped cream depends upon the clumping together of the fat particles. The best whipping cream did not give as large a volume as some other creams, but it had less liquid drain out of it after whipping.

Cream whips better with an increasing fat content up to 35 per cent. The cream with the higher fat content gives more particles for clumping and also increases the viscosity of the cream.

As the fat particles clump at the liquid/air interface, or within the liquid, the increased rigidity they give the foam permits inclusion of more air bubbles and extension of the films with the result that the dryness of the foam is increased. Larger fat particles clump more readily and thus form the structural support offered by the fat more easily. This offers an explanation of why cream from milk containing larger fat particles, milk from Jersey and Guernsey breeds, whips more quickly than cream containing smaller fat particles, milk from other breeds.

Aging improves the whipping qualities of cream. The viscosity increases with aging. As a general rule, treatment that increases the viscosity increases the whipping properties. Pasteurization tends to reduce the whipping quality of cream.

Babcock has found that the best whipping is obtained at a temperature of 45°F. or lower. At this temperature agitation favours the clumping together of the fat particles. At high temperatures, both the higher temperature and the agitation increase the dispersion of the fat. Above 50°F. the decrease in stiffness of whipped cream is in direct ratio to increase in temperature, so that 30-per cent cream will not whip at 72°F.

Babcock found acidity up to 0.3 per cent, at which sour taste is evident, had no effect on whipping quality. If acid was added in excess of 0.3 per cent, whipping quality improved, whether added to fresh or aged cream, when the amount added began to curdle the cream.

The addition of sugar to cream either before or after whipping was found by Babcock to decrease the stiffness of the cream. For each 2 tea-spoons added to 100 cc. (about 1/2 cup) the stiffness decreased four points on the stiffness scale. Adding the sugar before whipping the cream decreased the volume obtained and increased the whipping time.

The denaturation or coagulation of the protein film at the air/cream interface, increases the stiffness of the whipped cream. Since colloidal reactions require time, it is a better practise to add sugar to whipped cream after, rather than before or during the whipping process. By this procedure denaturation is more complete, which offers an explanation of why the volume of the cream is less affected by the addition of the sugar last. The addition of sugar lessens the stiffness and decreases the volume of the whipped cream because it either prevents denaturation and/or peptizes the protein film.

Reaction of Milk

Freshly secreted milk is nearly neutral to litmus. The reaction varies slightly but has an approximate pH of 6.6. The freshly secreted milk contains carbon dioxide. The amount of this gas in the milk decreases during milking and the subsequent handling of the milk, while the percentage of oxygen and nitrogen increase. For this reason the titratable acidity decreases for a time in milk exposed to the air. Confined milk does not show as great a decrease in titratable acidity as the exposed milk, for the percentage of carbon dioxide lost is smaller.

Effect of heating milk on acidity. When milk is heated at the boiling point or at temperatures above or near the boiling point the titratable acidity at first decreases owing to the loss of carbon dioxide, and then increases. Whittier and Benton report that the hydrogen-ion concentration increases continuously. They find the hydrogen-ion increase and the later increase in titratable acidity is due to the formation of acids from constituents of the milk. The amount of acid produced depends upon the time and temperature of heating, a greater amount of acid being produced with a longer heating period and with higher temperatures. From their experiments they conclude that the acid is produced from the lactose of the milk. They have shown that, the greater the concentration of lactose present, the greater the amount of acid formed at a definite temperature and for a definite time.

Coagulation of Milk

Under certain conditions, the addition of alcohol as well as the application of heat may cause coagulation of milk. Milk may be coagulated by the addition of rennin or by bringing the acidity of the milk to the isoelectric point of the casein. When milk is combined with other foods, the salt content of the food or the tannin content of the food may be factors that aid coagulation.

Alcohol Coagulation: The addition of 70 per cent alcohol to milk may cause coagulation. As the pH of the milk decreases, it becomes susceptible to alcohol precipitation, though this varies with different milks. Usually the milk is precipitated by alcohol while it is still stable to heat and sterilizing temperatures. Some freshly secreted milk is coagulated by alcohol, but this milk is usually abnormal in some way. About 2 cc. of alcohol are added to an equal quantity of milk for the test. Casein is precipitated by alcohol as calcium caseinate, calcium is not released as by acid coagulation.

Rennet Coagulation: Milk may be coagulated by the addition of rennet. Rennet is an extract that is usually obtained from the inner lining of the stomachs of calves and lambs. The rennet contains an enzyme called rennase or rennin. The clotting of the milk is generally believed to be the direct action of the rennin on the casein. But the manner in which these changes is produced is not fully understood. A very small amount of rennin is capable of coagulating a large amount of milk. At favourable or optimum hydrogen-ion concentrations for clotting, 1 part of the fairly pure enzyme preparation is able to coagulate 3,000,000 or more parts of milk.

Mechanism of Clotting: It is usually stated that the casein is changed to paracasein by the action of the rennin. It is also often stated that the clotting is brought about in two steps, the first being the action of rennin on the casein and the second the precipitation of the changed casein. Rogers reviews the many theories of rennin coagulation. Some investigators claim the changes are purely chemical; others maintain the rennin affects only the physical state of the calcium caseinate. However, if the change can be explained on the basis of colloid chemistry, it is probable that absorption and the electric charge play an important role in the process. Rogers states that Hammarsten regards casein in milk as a calcium caseinate-calcium phosphate complex. "As a matter of fact the compound called calcium caseinate is most probably a true calcium phosphocaseinate, if, as seems likely, the second and third hydrogens of the orthophosphoric acid esterfied with certain of the amino acids in the casein molecule react with calcium. The correct conception of the term 'calcium phosphocaseinate,' as it is now commonly employed, is that of a colloidal calcium phosphate (or phosphates) sol protected by a calcium caseinate (or caseinates) sol in a manner as yet imperfectly understood." The stabilization of sols is best explained by the theory of Helmholtz, i.e., each colloidal particle is surrounded by an electrical double layer. "In the case of negatively charged sols, in which class calcium caseinate and calcium paracaseinate evidently fall, the outer layer consists of hydrogen ions. If these are replaced by a sufficient number of positively charged ions of higher charge, e.g., calcium ions carrying two positive charges"; or, in other words, if these ions are more strongly adsorbed than the hydrogen ions, the colloid particle will readily precipitate, the rate of clotting being determined by the rate of replacement.

Richardson and Palmer state that rennin itself may reduce the charge of the calcium caseinate micelle and thus reduce the stability of the casein sol. They found indications that the isoelectric point of rennin is about pH 6.9 to 7.0. Above this pH the rennin is negatively

Fluid Milk Processing 185

charged and below pH 6.9 it is positively charged. They found that rennin lowered the electro-phoretic velocity of calcium caseinate and calcium phosphocaseinate micelles when the casein sol was negatively charged and the rennin was positively charged, but not when the rennin was negatively charged (above its isoelectric point, pH 6.9 to 7.0). From this evidence and from the fact that paracaseinate micelles are not affected by rennin, which agrees with the fact that casein once coagulated by rennin has lost its sensitiveness to this enzyme, they suggest that rennin acts by sensitizing the casein by a preliminary reduction of the electric charge on the casein micelles.

During the clotting of the milk, aside from the consistency of the milk, there is little change in its physical properties. The hydrogen-ion concentration does not change during the clotting process.

Factors affecting action of rennin. Several factors influence the activity of the rennin in bringing about coagulation. These may be listed as follows: (1) temperature for rennin action; (2) heating the milk before the addition of rennin; (3) hydrogen-ion concentration; (4) concentration of casein, calcium, and phosphate ion; (5) character of cations used for coagulation.

Temperature for rennin action. The optimum coagulation by calf rennin is about 40° to 42°C. Below this temperature coagulation is less rapid and no clotting occurs below 10° to 15°C. Also no clotting occurs above 60° to 65°C. The clot is softer at low temperatures and tougher and stringy at high temperatures. By optimum is meant the temperature at which coagulation takes place most rapidly for a definite concentration of rennin and milk. Effect of previously boiling the milk upon rennin coagulation. If milk is boiled and then cooled before the rennin is added, the rate of coagulation is retarded and a much softer, more flocculent clot is obtained. Pasteurization also affects the rate of coagulation of the milk and the type of clot formed by rennin but not to the extent that boiling does.

Richardson and Palmer found by electrokinetic evidence that heat increased the electric charge on the casein micelles or the cataphoretic velocity of the casein solution. The fact that rennin does not form as firm a clot with milk that has been previously heated indicates that rennin reduces the charge on the casein particles but not sufficiently to form a firm clot. This offers a colloidal explanation of why the addition of active cations (as calcium chloride) to heated milk causes the rennin to coagulate the milk normally.

Hydrogen-ion Concentration: The reaction of the milk affects the rapidity of coagulation and the character of the curd formed. Ordinarily when the reaction of the milk is alkaline coagulation does not occur. This is shown by the addition of a small amount of soda to milk before the addition of junket. The optimum hydrogen-ion concentration for rennin activity has been reported to lie in the zone between pH 5.99 and 6.40.

Character of Cations: In addition to rennin, cations are necessary to bring about coagulation of milk. Because casein and calcium are so closely involved in milk, the cation calcium is important in bringing about coagulation. Hence, Rogers states that it is to be expected that the concentration of both casein and calcium markedly affect both the rate of coagulation and the character of the clot. If milk is diluted with sufficient water, clotting is both delayed and incomplete, the clot being soft. If calcium chloride is added to the water, diluted milk clotting properties are restored, which suggests that the concentration of calcium ions is more important than that of the casein ions.

Rogers states that any metallic ion can replace the calcium in coagulation. However, it is generally accepted that the sodium and potassium salts of paracasein are soluble. Monovalent ions are less effective than divalent ones in replacing the calcium. Rogers reports that all monovalent ions did not bring about coagulation in some instances. The divalent ions were not all equally effective, calcium and barium being more efficient than magnesium.

Sugar: Sugar tends to prevent the coagulation of milk by rennin.

Coagulation of milk by acid. Kruyt states that there are some proteins that are not sufficiently hydrated to be stable by hydration alone. He cites casein as an example of a protein "which can exist either in acid or in an alkaline solution, but does not dissolve in water, with the consequence that the sol ordinarily flocculates when neutralized." Either the acid produced during fermentation or acids added to milk precipitate the casein. The casein is least soluble at its isoelectric point pH 4.6. If enough acid is added to lower the pH below 4.6, casein salts, such as casein chloride or casein lactate, are formed. If these salts are soluble, the casein goes into solution. Hence the largest yield of precipitated casein is near the isoelectric point.

Fermentation of Milk: Fermentation, or the production of lactic acid from lactose by bacteria, takes place in milk that is allowed to stand under favourable conditions. Rogers states that true lactic acid fermentation is brought about by the Streptococcus lactis and certain other organisms, lactic acid being the principal end-product, other products being present in only small amounts. In mixed lactic acid fermentation, or when other organisms in addition to S. lactis are present, the end-products may include acetic, propionic, lactic, succinic, formic, and butyric acids, carbon dioxide, hydrogen, acetone, and ethanol. As fermentation increases, an acidity is reached at which the action of most bacteria is suppressed. When fermentation is checked at pH 4.8 to 5.0, the bacteria consists chiefly of Streptococcus lactis.

The rate of fermentation depends chiefly upon the temperature at which the milk is held. At low temperatures, on account of retardation of bacterial action, it takes place slowly. Rogers states that fermented milk, allowed to stand at a fairly high temperature, undergoes a second lactic acid fermentation brought about by the Lactobacillus bulgaricus organisms. Some of these types of bacteria form a high percentage of acid and the hydrogen-ion concentration may reach pH 3.23.

Changes occurring during acid precipitation. During fermentation chemical and physical changes occur in the milk. The flavour becomes acid. The calcium caseinate is changed to casein. During this process calcium is split off and forms soluble calcium lactate. In addition some dicalcium phosphate is converted into monocalcium phosphate. Curdling or clotting occurs when the acidity reaches about pH 5.3. During the clotting process the hydrogen-ion concentration does not increase. Milk clotted by fermentation is often called clabbered milk. Its flavour and aroma may vary, depending upon the types of bacteria producing the fermentation. Fermented milk may be used for drinking, for cooking, and for cottage cheese.

Cheese, such as cottage cheese, when clotted by acid coagulation, loses a large proportion of its calcium. The calcium salts become soluble more rapidly than the phosphorus; hence a larger proportion of the calcium than of the phosphorus is lost in the whey. Casein precipitated by rennin retains most of its insoluble salts, hence has a larger proportion of calcium than the acid precipitated casein.

Heat Coagulation: The term heat coagulation refers to the so-called "denaturation" of the protein, by which it is rendered insoluble.

Lactalbumin. Lactalbumin has temperatures for heat coagulation similar to that of egg albumin. The lactalbumin forms a flocculent precipitate, whereas egg albumin forms a firm coagulum. Rupp has reported the following amount of lactalbumin coagulated when heated for 30 minutes.

Temperature,°c.	Albumin rendered insoluble, per cent
62.8	0.00
65.6	5.75
68.3	12.75
71.1	30.78

Casein. Casein is not coagulated by heat at ordinary temperatures or when heated for short periods, though the heating may alter the casein. Rogers states that it is necessary to heat milk about 12 hours at 100°C. to bring about coagulation. It takes approximately 1 hour at 135°C. and approximately 3 minutes at 155°C. The time and temperature vary somewhat with different milks.

The rate of coagulation depends upon the concentration of the casein as well as the time and the temperature of heating. Rogers states that evaporated milk, containing twice the concentration of solids-not-fat in normal milk, and thus a higher concentration of casein, requires about 60 minutes for coagulation at 114.5°C, 10 minutes at 131 °C, and 7500 minutes at 80°C.

Rogers and Palmer both state that, in the evaporated-milk industry, the forewarming of milk prior to processing increases its stability to heat. "Rapid improvement in resistance to heat coagulation results in increase in temperature for prewarming up to 90° to 100°C. Above 90°C. the change is very small, but in some cases can be effected with increases in temperature up to 120°C. for 10-minute periods of forewarming. When time is chosen as the variable, improvement may be noted with increases in the time up to 30 minutes at a temperature of 95°C. At higher temperatures the same improvement may be effected in shorter periods of time."

Fat: Rogers states that fat particles in relatively large aggregates may act as nuclei about which coagulation of the casein can proceed. In un-homogenized milk the fat affects the coagulation time and temperature but slightly. But when a milk of higher fat content is homogenized the fat clumps may act as nuclei about which the casein may gather during heating. With increase in homogenization pressure as well as fat content, other conditions being the same, a marked decrease in stability to heat is noted. In homogenized milk it was found that the maximum stability to heat coagulation occurs if

Fluid Milk Processing 189

homogenization is carried out at 80°. Rogers adds that the feathering of some homogenized cream when added to coffee may be caused by using too high homogenization pressure, thus reducing the stability of the cream to heat.

The role of salts in heat coagulation of milk. In heat coagulation of milk, the milk salts play an important role, for the salt equilibrium is altered by heat. When milk is boiled precipitation of part of the calcium phosphate occurs. Sommer and Hart have concluded that salts are the main factor in heat coagulation of fresh milk.

Electrolytes have a marked effect upon the stability of colloids. In precipitating a hydrophilic colloid divalent and trivalent ions are generally more effective than monovalent ones. In the milk are found the monovalent cations, sodium and potassium; the monovalent anion, chlorine; the divalent cations, calcium and magnesium; and the trivalent anions, phosphate and citrate. Sommer and Hart concluded that the coagulation of milk on heating may be due to an excess or a deficiency of calcium and magnesium.

They explain this as follows. "The casein of the milk is most stable with regard to heat when it is in combination with the calcium. If the calcium combined with the casein is above or below this optimum, the casein is not in its most stable condition. The calcium of the milk distributes itself between the casein, citrates, and phosphates chiefly.

If the milk is high in citrate and phosphate content, more calcium is necessary in order that the casein may retain its optimum calcium content after competing with the citrates and phosphates. If the milk is high in calcium there may not be sufficient citrates and phosphates to compete with the casein to lower its calcium content to the optimum. In such cases the addition of citrates or phosphates makes the casein more stable by reducing its calcium content. The magnesium functions by replacing the calcium in the citrates and phosphates."

Heat coagulation of casein endothermic. Leighton and Mudge have shown that an endothermic reaction accompanies the appearance of visible curds when milk is coagulated by heat. This is accompanied by precipitation of calcium and magnesium as phosphate and citrates. A similar reaction occurs in custard.

In cooking custard, the ingredients of which are milk, egg, and sugar, the temperature drops or does not rise for a period of time during coagulation or setting of the custard, a condition particularly noticeable just before curdling takes place.

Coagulation of milk by cooking meat or vegetables in it. Fresh milk is seldom coagulated by heating for home use. The temperature attained in ordinary heating is not great enough to cause coagulation, nor is the milk heated for the long period required for coagulation at boiling temperatures. But with the addition of other foods to milk in food preparation, coagulation often occurs with a very short period of heating. One of the factors in this coagulation is undoubtedly the salt content of the food added to the milk as well as the salt content of the milk.

The balance of the milk salts for greatest stability may be upset and coagulation occurs when the food is heated in the milk. Cooking of meat in milk. Ham is often baked in milk. Sometimes pork chops are floured, seared in fat, and then baked in milk. Other meats and fish are sometimes baked or cooked in milk. Often curdling of the milk occurs, and the appearance of the meat, owing to adherence of curds of milk, is not attractive. Thus it is desirable to prevent curdling. Among the causes of curdling are the temperature at which the meat is cooked, the salt content of the food cooked in the milk, the manner in which the milk is added to the meat, and the reaction of the milk.

The higher the temperature at which the meat is cooked the greater the tendency to curdle. Larger curds are also formed at higher temperatures. The temperature to which milk must be heated to bring about curdling is high, so this cooking temperature alone is not sufficient to bring about the coagulation. Even when the meat is cooked in an oven at a high temperature, the liquid portion does not reach a higher temperature than boiling. It is possible that the acidity is increased during cooking, but the resulting pH of the meat-milk broth is changed very little from that of the original milk or from that of the meat broth when the meat is cooked in water. Thus it seems that the heating and the acidity developed during cooking do not alone bring about coagulation. The salt content of the food cooked in the milk probably influences the coagulation, and this combined with the heating, the temperature of heating, the acidity developed, and the altering of the casein by heat are sufficient to cause curdling. Sodium chloride has some effect, for curdling is more likely to occur when the meat cooked in the milk is a cured or salted one than when fresh meat is used.

Curdling may be prevented by the addition of soda, about 1/16 teaspoon per cup of milk. From this it appears that the reaction has some part in the coagulation. The soda may combine with other salts that tend to bring about coagulation or the coagulation may be prevented by the slightly alkaline reaction. A slightly alkaline reaction also prevents coagulation by rennin and by fermentation.

If a portion of the milk is added to the meat when cooking is first started and the rest of the cold milk added gradually to the meat during the cooking period, curdling is less likely to occur. The addition of acid foods, such as prepared mustard, which may contain vinegar, or apples and pears to be baked with the meat, would tend to increase the tendency to curdle. Evaporated milk has less tendency to curdle than fresh milk, which may be due to the previous heating.

Cooking of vegetables in milk. Milk usually does not curdle when cabbage, chard, spinach, or cauliflower is cooked in it. But it is likely to curdle when asparagus, string beans, peas, and carrots are cooked in it. Asparagus usually curdles the milk after a few minutes of cooking. There are several factors that may aid in bringing about coagulation of the milk. The slight acidity of some vegetables combined with the heating of the milk may tend to bring about coagulation, but the acidity is not usually great enough, nor the boiling temperature high enough, nor the boiling long enough continued for these factors to be very important. The salt and the tannin contents of the vegetables are probably the principal causes of coagulation.

Some vegetables contain larger amounts of tannin than others. Tannin is a dehydrating agent and brings about denaturation of hydrophilic sols, like gelatin, starch, and agar-agar. After denaturation the hydrophile is sensitive to small amounts of electrolytes and precipitation occurs readily. Kruyt states that tannins do not bring about dehydration of the protein in an alkaline medium.

Hence the addition of soda in small amounts to the milk in which the vegetable is cooked prevents coagulation of the milk. Tannins lower the surface tension, which results in foaming of vegetables containing tannin when they are cooked in water. It is rather interesting that the vegetables that usually foam the most when cooked in water are the ones that have the greatest tendency to coagulate milk.

Tomato soup. When tomatoes are combined with milk to make cream of tomato soup, coagulation may occur. The acidity of tomatoes varies somewhat, but is about pH 4.4 to 4.6. If the amount of tomato added to the milk is great enough to lower the pH of the mixed milk and tomato to 4.8 to 4.6, the casein is precipitated without heating. This may happen if the milk is already fairly acid.

Since a longer time of heating milk increases the tendency to curdle it is preferable to heat tomato soup for only a short time. Heating slowly also increases the tendency to curdle, which may be due to the longer time required. In Experiment 55 in the laboratory outline several different methods of combining the tomato juice and milk are given. The tomato is usually added to the milk by stirring, for in this way the milk is diluted with a smaller amount of acid substance during the first part of the mixing. There is less tendency to curdle when the hot tomato is added to cold milk, than when the cold tomato is added to hot milk. Probably the slight denaturation brought about by heating the milk may partially account for this. Occasionally some milk is acid enough and the tomato is acid enough to cause curdling with all methods of combining unless soda is added. At other times curdling does not occur with any method of combining as outlined in this experiment.

Fruits and milk. When cream is added to fruit, clotting often occurs. This is usually due to the acidity of the fruit, but may also be due to an enzyme in it. Raw pineapple contains an enzyme, bromelin, that brings about clotting of milk. However, the pineapple juice not only brings about clotting but also peptization, for after a time the clot formed is less firm and the flavour is similar to that of peptized meat.

Boiling and Heating of Milk

The physical and chemical properties of the constituents of milk account for the behaviour of milk during its use in food preparation. Thus substances that lower surface tension become concentrated in the liquid/air interface. Proteins lower the surface tension of aqueous sols, hence accumulate in the surface. When milk is heated in an open pan, a scum or skin forms over the surface of the milk. At first this skin is rather thin and mobile but is gradually altered so that it becomes tenacious and tough enough to be removed with a stirring rod or spoon. This scum has been said to contain coagulated albumin and globulin. Tinkler and Masters state that if the scum is removed as formed, the total amount of protein that can be removed exceeds the total amount of albumin and globulin in the milk. When foods are cooked in milk the milk not only foams readily but the scum tends

Fluid Milk Processing 193

to hold the steam formed in heating the milk; it is because of this that the milk "boils over" so readily.

Sugar Reactions with Proteins of Milk

Ramsay, Tracy, and Ruehe investigated the substitution of dextrose for sucrose in sweetened condensed skim milk. They found the objections to using dextrose were (1) a brown discoloration, (2) a physical thickening, and (3) crystallization of the dextrose during storage. The last objection could be remedied by using 50 per cent dextrose and 50 per cent sucrose. The progressive thickening during storage at high temperature was caused by action of the dextrose on the casein and albumin of the milk.

During this investigation they found additional evidence that sugars react with proteins. When dextrose, lactose, or levulose was heated with skim milk or freshly precipitated casein, a dark brown colour formed in the product. When the sugars were heated in distilled water solutions to 250°F. for 30 minutes no caramelization occurred. Neither did darkening occur when albumin or casein was heated in water solution. But when the milk, albumin, or casein was heated with lactose or dextrose, a brown discoloration occurred. As the temperature was raised the dextrose and casein became so firmly attached to each other that no amount of washing could remove the sugar. Most of the biruet action of the skim milk was lost. The results were explained on the basis that a protein-sugar complex of glucosidal nature was formed.

On heating amino acids with dextrose highly coloured products were formed, the reaction probably being a condensation of an amino acid with an aldehyde or ketone group of the sugar. The very stable linkage of the aldehyde group of the dextrose and the ketone group of the fructose in the sucrose molecule is cited to explain the failure of sucrose to form condensation products with casein, albumin, or amino acids. It was found that as the reaction became more alkaline the appearance of the brown colour was more rapid.

The increased alkalinity was said to favour the change of the sugar from a lactone to a free aldehyde form, the free aldehyde acting with the amino acid or -NH1 groups of the protein. If the pH was much above 7 the milk was almost black. Hence, the sugar used, the reaction, and temperature all influenced the development of the brown colour. The length of heating in connection with the temperature was important as relatively high temperatures for a short period gave only slight development of the brown colour. It is almost impossible to retain the natural colour of fresh milk in the condensed milk products, for some brown discoloration occurs in the unsweetened and sweetened product whether made from whole or skim milk.

Whittier and Benton had shown that the hydrogen-ion concentration increases at a rate which is the function of the lactose concentration and the time and temperature of heating. Or, in other words, when milk is heated for a sufficient time at high enough temperatures the lactose is decomposed with formation of acid products. Hence, when milk is heated with sucrose the increasing acidity inverts some of the sucrose to dextrose and levulose, with the development of a brownish colour. One example of this is in the cooking of caramels, more brown colour developing with long slow cooking of the sucrose and milk. Another instance where this is used to advantage is in making caramel pudding by boiling, in the can, sweetened condensed milk for three hours or longer. The can and contents are chilled. On opening the can it is found that the contents have developed the brown colour of caramelized products and are thickened to the consistency of a pudding. This combination of sugar with milk proteins to form a thickened product is interesting in view of the fact that sucrose, dextrose, and levu-lose prevent the heat coagulation of egg albumin.

The housewife also makes use of the effect of acid on sweetened condensed milk. If about 1/2 cup of lemon juice is stirred into the contents of a can (about 1 1/2 cups) of sweetened condensed milk, the mixture thickens to a consistency that can be used for a pudding or pie filling and may be thinned with water to a desired consistency. The explanation of the thickening lies in the action of the acid on the complex sugar-protein combination.

Cheese

Definition: The Food and Drug Administration defines cheese, in the regulatory announcements, as "a product made from curd obtained from the whole, partly skimmed, or skimmed milk of cows, or from milk of other animals with or without added cream, by coagulating with rennet, lactic acid, or other suitable enzyme or acid, and with or without further treatment of the separated curd by heat or pressure, or by means of ripening ferments, special molds, or seasoning."

Fluid Milk Processing　　　　　　　　　　　　　　　　　　　　　　　195

    Classification of cheese. Cheese may be classified in many ways as (1) method by which the curd is produced, i.e., acid or rennet coagulation, (2) source of the milk, from cow, sheep, or goat, and (3) the texture and consistency of the cheese, i.e., whether soft, semi-hard, or hard. Many other classifications might be used but none of them are entirely satisfactory. Doane and Lawson list and describe nearly 300 cheeses. They state there are probably about 18 distinct varieties of cheese.

    For purposes of discussion, Rogers classifies cheese as follows:

Soft

- Unripened Cottage Cream Neufchatel Ripened
- Ripened by molds
- Camembert
- Brie Ripened by bacteria
- Limberger
- Liderkranz.

Hard

- Semi-hard
- Ripened by molds
- Gorgonzola
- Roquefort
- Stilton Ripened by bacteria
- Brick
- Munster Very hard
- Without gas holes
- Cheddar
- Edam
- Gouda With gas holes
- Emmenthal
- Swiss
- Parmesan.

    Composition of cheese. From the standpoint of quantity the principal constituents of cheese are casein, fat, and water. In addition it contains various salts, and unless heated to pasteurization temperatures various organisms such as bacteria and molds. Different types of soft cheese may contain from 40 to 75 per cent of water, hence this type does not keep long. Hard-type cheeses usually average 30 to 40 per cent moisture. The soft types may contain from 13 to 21 per cent of protein and from 0.5 to 50 per cent of fat. Hard types contain from 20 to 45 per cent protein and 19 to 40 per cent fat.

Coagulation of Milk for Cheese: Coagulation may be brought about by rennet or acid. Rennet-formed curds are more elastic, the acid ones more sticky. In acid-formed curds more of the calcium salts are split off from casein, forming calcium chloride which is soluble in the whey. Rennet-coagulated cheeses of cheddar types retain about 80 per cent of the calcium of milk, whereas soft cheeses retain about 20 per cent.

The temperature for coagulation varies with the type of cheese desired. In general, the lower the temperature the softer the curd. Curds formed at 21° to 25°C. are used for some soft cheeses. Cheddar cheese has a firmer curd and the milk is brought to 30°C. before the starter and rennet are added. Temperatures as high as 48°C. may be used for some cheese, the curd produced being distinctly tough and somewhat rubbery and elastic.

Making Cheese: The essential steps in making cheese are: blending the particular type of milk desired; bringing the milk to a definite temperature; adding lactic acid culture for types of cheese that need greater hydrogen-ion concentration when the rennet is added (acid cultures are added to cheddar types, but not to Swiss); adding vegetable colour, if cheese is to be yellow, omitting if cheese is to be American white; and adding the rennet. After coagulation the curds are cut to the definite size for the type of cheese desired.

Small curds retain less moisture within the curd but the whey does not drain so well from the curd. The next step is stirring the curd gently to facilitate draining of the whey. The curd is then ditched, salted, put in molds lined with cloth, and pressed into definite shapes as Longhorns, Prints, Daisies, Flats, Twins, and Cheddars. After being pressed the cheese may be soaked in salt brine or dry salt may be rubbed on the surface. Sometimes no additional salting occurs.

Soft unripened cheese is not cured; but after being pressed or moulded other types are placed on shelves in caves or specially constructed curing rooms to ripen. In the latter ventilation, humidity, and temperature may be carefully controlled according to the type of cheese. The curing period varies for different types of cheese and for the same type. For example, Cheddar may be cured from 2 or 3 months to 2 years. With longer curing a sharper, richer, and fuller flavour is developed.

Cheese, after being cured, is often blended for uniform flavour, texture, and body.

Secondary Heating of the Curd: A secondary heating of the curd is necessary with most hard and semi-hard cheeses. Making Emmenthal involves heating to about 55° to 58°C. This heating hastens the driving of the whey from the curd, changes its texture, and often alters the bacterial flora. The heating at high temperatures decreases the moisture content and rennet action is checked if not wholly stopped. Various physical changes take place during this period, the curd becoming tough, firmer, and rubbery. In Swiss and Parmesan cheese it also acquires plasticity.

Ripening of Cheese: In the process of ripening chemical and physical changes occur in the cheese. It loses its tough, rubbery qualities and becomes soft and mellow, sometimes almost crumbly. During this change as much as 50 per cent of the nitrogenous constituents may be converted to soluble forms, though the average for hard cheese is 30 per cent. These changes not only alter the texture and flavour, but also alter the cooking quality of the cheese, the increased solubility of the proteins increasing the ease with which the cheese may be blended with eggs, milk, and white sauce.

Ripening is slower at lower temperatures and more rapid at higher ones. Not only enzymes of the milk, if the milk has not been heated to a temperature to destroy the enzymes, but bacteria aid in ripening of the cheese and hydrolysis of the proteins. Some bacteria, such as lactic acid, produce enzymes that split the protein. More hydrolysis occurs in the softer centre of hard cheese than near the rind.

Salting affects the rate of ripening by delaying bacterial growth, the proteins of cheese with more salt becoming soluble at a slower rate. Salt penetrates slowly from the rind to the centre and aids in drying the cheese. Changes in the fat in the interior of most cheese are usually negligible.

For the growth of molds and aerobic bacteria, holes must be punched in the cheese to allow oxygen from the air to penetrate. In the early stages of ripening Emmenthal and Swiss cheeses are soft and become elastic. It is during this stage that the holes or "eyes" are formed from production of gas, principally carbon dioxide, if ripening is normal, but with more hydrogen in abnormal or early ripening. If the cheese becomes too firm before the formation of holes is complete, checks and cracks appear in the cheese.

Cheddar Cheese in Cans: The Bureau of Animal Industry (Rogers) has announced a practical method of canning unripened Cheddar cheese. By this method a one-way or check valve, which holds perfectly against external pressure but with internal pressure allows gases formed during ripening of the cheese to escape, is inserted in the lid of the can.

Cheddar cheese has always been pressed in cylindrical forms of varying sizes, but in general rather large. When these large cheeses are cut they lose moisture, so the cut surface dries rapidly. In addition, if the cheese is well ripened, loss occurs through crumbling. In packing cheese in cans, the cheese, after pressing, is cut into the desired shape. Since hydrogen sulfide is often liberated during ripening of cheese, it is preferable to wrap the cheese in parchment and it is necessary to use a lacquered can, for the hydrogen sulfide tends to form a black product with metals such as iron, copper, or lead.

Processed Cheese: Rogers states that before the development of the can in which Cheddar cheese may be ripened, the "only commercial method for putting Cheddar cheese into a more attractive and convenient form is the one known as processing. After the rind is removed, the cheese is ground, a small quantity of water and an emulsifier, usually sodium citrate, are added, and the mass is heated with constant stirring until it becomes fluid. The emulsion is run into forms, which in many cases are boxes lined with tinfoil, in "which it is sold.

The cheese hardens quickly and, as the wrapping adheres closely, there is no trouble from molds. Moreover, as the temperature is high enough to constitute pasteurization, most of the bacteria are killed and the enzymes destroyed, so that ripening is stopped. In this process, much of the original character of the cheese is lost; but, in spite of this objection, the advantage of the package is so great that a large part, possibly one-third, of all the cheese made in the United States is sold in this form."

Templeton and Sommer have investigated various salts that may be used as emulsifiers in processed cheese. They state the purpose of the salt is to prevent separation of the fat from the cheese and at the same time give the finished product the desired body and texture. They quote Habicht as stating that an alkaline monovalent cation combined with a polyvalent anion, such as sodium citrate, is the ideal emulsifying salt. The physico-chemical explanation is as follows: There is partial saponification between the cation (sodium, if sodium citrate is used) and the fatty acids. The soaps formed are good emulsifiers. In addition the anion, which is a solvent for casein, combines with the casein of cheese so that a film of casein surrounds each fat globule, thus emulsifying it and preventing its escape from the mass. Later we find that the citrate ion is also a good peptizer of egg and flour proteins.

Loaf Cheese: Rogers states that blending is used extensively for Cheddar and Swiss cheese. In this process the cheese is ground and heated in steam-jacketed kettles, 60° to 70°, and then poured into molds. In the initial heating separation of the fat occurs; but with longer heating the casein becomes plastic and stringy and encloses the fat. Further agitation causes the mass to lose its plasticity and become the consistency of heavy cream. At this time it is poured into the molds.

The plasticity of the cheese is an important part of the process. Once the plasticity is broken it is almost impossible to restore it. The method of manufacture, the degree of ripening, the acidity of the cheese, and possibly other factors influence the degree of plasticity attainable in the heated cheese and the length of time the mass will remain plastic. Sodium and ammonia seem important in the emulsification of the product.

Cheese Spreads: The term cheese spread may be applied to any packaged form of cheese that can be easily spread with a knife at ordinary room temperature. Templeton and Sommer name the types on the market as: (1) cream cheese, mixed with pickles, olives, etc., (2) processed cheese of such age and moisture content as to be "spready," and (3) processed cheese with concentrated whey or skim milk powder added and of such fat and moisture content that the mix will spread easily.

They say that, since the composition is quite different from cheese, as defined for Food and Drug regulations, the product cannot be sold as cheese. Actually they are sold as food products under proprietary trade names. The desirable spreading qualities may be due to the moisture content or the fat content or both.

The use of cheese in cooked products. All of the factors that affect the plasticity of the cheese when heated for blending, i.e., the degree of ripening, the acidity, and method of manufacture, also affect its blending properties with other ingredients in such dishes as rarebit, cheese souffle, and macaroni and cheese. To these factors may be added the extent of drying. For the cut or grated surface of cheese may dry rather extensively. Hence, the protein in the surface area really needs soaking for hydration before it will blend with other ingredients, or it may entirely lose its plasticity.

Cream cheese may be combined with eggs, sugar, etc., for cheese cake or similar cooked dishes. But, in general, whether in its original state after curing, or processed, the Cheddar type is the cheese usually combined with cooked products.

The cheese is combined with white sauce or eggs at low temperatures and by stirring. The temperature should be as low or lower than that used for blending, 40° to 50°C. often being preferable to 60° to 70°C. As the protein becomes plastic the fat exudes. Stirring aids in emulsifying this fat with white sauce and casein of the cheese.

Cheese Souffle: A colleague, Plagge, suggested a good method of combining the ingredients for cheese souffle. The beaten egg yolks were added to the white sauce before the grated cheese, because the addition of the egg yolks cooled the mixture to a greater extent before the cheese was added. However, another advantage of this order of mixing is that the egg yolk aids in emulsifying the fat of the cheese. For the same reason beating with a rotary egg beater as the cheese softens is a good method of blending the cheese with the white sauce, since it more efficiently divides the cheese, thus increasing the surface area for emulsification.

Processed cheese usually combines particularly well with white sauce and egg yolk, because of its added water content and the emulsification of the fat.

6

Heat Treatments and Pasteurization

The Purpose of Pasteurization

1. To increase milk safety for the consumer by destroying disease causing microorganisms (pathogens) that may be present in milk.
2. To increase keeping the quality of milk products by destroying spoilage microorganisms and enzymes that contribute to the reduced quality and shelf life of milk.

Pasteurization Conditions

Minimum pasteurization requirements for milk products, and are based on regulations outlined in the Grade A Pasteurized Milk Ordinance (PMO). These conditions were determined to be the minimum processing conditions needed to kill *Coxiella burnetii*, the organism that causes Q fever in humans, which is the most heat resistant pathogen currently recognized in milk. Milk can be pasteurized using processing times and temperatures greater than the required minimums.

Pasteurization can be done as a batch or a continuous process. A vat pasteurizer consists of a temperature-controlled, closed vat. The milk is pumped into the vat, the milk is heated to the appropriate temperature and held at that temperature for the appropriate time and then cooled. The cooled milk is then pumped out of the vat to the rest of the processing line, for example to the bottling station or cheese vat. Batch pasteurization is still used in some smaller processing plants. The most common process used for fluid milk is the continuous process. The milk is pumped from the raw milk silo to a holding tank that feeds into the continous pasteurization system. The milk continuously flows from the tank through a series of thin plates that heat up the milk to the appropriate temperature. The milk flow system is set up to make sure that the milk stays at the pasteurization temperature for the appropriate time before it flows through the cooling area of the pasteurizer. The cooled milk then flows to the rest of the processing line, for example to the bottling station. There are several options for temperatures and times available for continuous processing of refrigerated fluid milk. Although processing conditions are defined for temperatures above 200°F, they are rarely used because they can impart an undesirable cooked flavour to milk.

History of Pasteurization

The process of heating or boiling milk for health benefits has been recognized since the early 1800s and was used to reduce milkborne illness and mortality in infants in the late 1800s. As society industrialized around the turn of the 20th century, increased milk production and distribution led to outbreaks of milkborne diseases.

Common milkborne illnesses during that time were typhoid fever, scarlet fever, septic sore throat, diptheria, and diarrheal diseases. These illnesses were virtually eliminated with the commercial implementation of pasteurization, in combination with improved management practices on dairy farms. In 1938, milk products were the source of 25% of all food and waterborne illnesses that were traced to sources, but now they account for far less than 1% of all food and waterborne illnesses.

Pasteurization is the process of heating a liquid to below the boiling point to destroy microorganisms. It was developed by Louis Pasteur in 1864 to improve the keeping qualities of wine. Commercial pasteurization of milk began in the late 1800s in Europe and in the early 1900s in the United States. Pasteurization became mandatory for all milk sold within the city of Chicago in 1908, and in 1947 Michigan became the first state to require that all milk for sale within the state be pasteurized. In 1924 the U.S. Public Health Service developed the Standard Milk Ordinance to assist states with voluntary pasteurization programs. The Grade A Pasteurized Milk Ordinance (PMO), as it is now called, is administered by the U.S.

Departments of Health and Human Services and Public Health, and the Food and Drug Administration and defines practices relating to milk parlour and processing plant design, milking practices, milk handling, sanitation, and standards for the pasteurization of Grade A milk products. Each state still regulates milk processing within their own state but dairy products must meet the regulations stated in the PMO for products that will enter interstate commerce.

Fluid Milk Production

Fluid Milk Definitions

Fluid milk is an industry term for milk processed for beverage use. Milk, as defined by the U.S. Code of Federal Regulations (CFR), 21 CFR 131.110, is: "the lacteal secretion, practically free from colostrum, obtained from the complete milking of one or more healthy cows. Milk that is in its final package form for beverage use shall have been pasteurized or ultrapasteurized, and shall contain not less than 8.25% solids and not less than 3.25% milk fat. Milk may have been adjusted by separating part of the milkfat therefrom, or by adding thereto cream, dry whole milk, skim milk, or nonfat dry milk. Milk may be homogenized." Milk solids are the non-water components of milk – protein, lactose, and minerals. Sometimes the combination of protein, lactose and minerals is called the solids not fat content, and when the fat is included it is called total solids content.

Although the CFR states that milk is obtained from cows, the production of milk from other dairy animals in the U.S. (goats, sheep, and water buffalo) also is covered in the Grade A Pasteurized Milk Ordinance (PMO).

Standardization

The fat content of milk varies with species (cow, sheep, goat, water buffalo), animal breed, feed, stage of lactation, and other factors. In order to provide the consumer with a consistent product, most milk in the U.S. is standardized.

To achieve standardization, milk is processed through centrifugal separators to create a skim portion and a cream portion of the milk. Separation produces a skim portion that is less than 0.01% fat and a cream portion that is usually 40% fat, although the desired fat content of the cream portion can be controlled by changing settings on the separator. The cream portion is then added back to the skim portion to yield the desired fat content for the product. Common products are whole milk (3.25% fat), 2% and 1% fat milk, and skim milk (< 0.1% fat).

Pasteurization

The majority of U.S. fluid milk is pasteurized using a high temperature short time (HTST) continuous process of at least 161°F (71.6°C) for 15 seconds. These conditions provide fresh tasting milk that meets the requirements for consumer safety. Higher heat processes, such as ultrapasteurization or aseptic processing, are used to extend the shelf life of refrigerated products or allow for storage at room temperature, respectively, but may impart a cooked flavour to the milk.

Homogenization

The fat in milk is secreted by the cow in globules of non-uniform size, ranging from 0.20 to 2.0 µm. The non-uniform size of the globules causes them to float, or cream, to the top of the container. Milk that is not homogenized is sometimes referred to as "creamline" milk. Pasteurized milk does not necessarily need to be homogenized. However, homogenized milk should be pasteurized to inactivate native enzymes that deteriorate fat (lipases) and cause rancidity, which results in off-flavours and reduced shelf life in milk.

The purpose of homogenization is to reduce the milk fat globules size to less than 1.0 µm which allows them to stay evenly distributed in milk. Homogenization is a high pressure process that forces milk at a high velocity through a small orifice to break up the globules. The result of homogenization is the creation of many more fat globules of a smaller size. The native milk fat globules are covered in a protein membrane that stabilizes the fat phase in the aqueous (water) phase of milk. Although the milk fat globule membrane is disrupted during the homogenization process, it spontaneously migrates back to the fat globules after homogenization. The new globules created during homogenization are spontaneously coated by proteins in the skim phase from the original milk fat globules.

Vitamin Fortification

Fluid milk is often fortified with vitamin A and vitamin D. The package label must declare when milk is fortified.

Whole milk is considered a good source of vitamin A. Vitamin A is a fat soluble vitamin that is found in the fat phase of milk. The vitamin A content that occurs naturally in 2%, 1% and skim milk is less than in whole milk because of the lower fat levels. Nutritional concerns about consumption of lower fat milk in the late 1970s led to the required fortification of vitamin A in lower fat milks. To achieve the nutritional equivalence of whole milk, lower fat milks should be fortified to 300 IU vitamin A per 8 oz serving. The FDA encourages fortification to a level of 500 IU of vitamin A per 8 oz serving, which is 10 % of the recommended daily allowance (RDA).

Vitamin D is a fat soluble vitamin that occurs naturally in milk but at low levels. Because milk is not considered an important natural source of vitamin D in the diet, vitamin D fortification is voluntary. Fortification of milk with vitamin D began in the U.S. in the 1930s to reduce the incidence of rickets in children. Although rickets is not currently a major concern in the U.S., adequate vitamin D is necessary for human health. Vitamin D helps with calcium absorption, has an important role in bone health, and has a protective effect in cancer. Milk may be fortified with vitamin D to a level of 100 IU per 8 oz serving, which is 25% of the RDA.

Specialty Milk Beverages

The dairy industry has developed specialty fluid milk beverages to meet the diverse nutritional needs of consumers. Lactose-reduced and lactose-free milk, and acidophilus milk were developed for people with lactose intolerance (maldigestion). Lactose-reduced and lactose-free milk are processed, prior to packaging, with the lactase enzyme to separate lactose into its component sugars, glucose and galactose. Acidophilus milk contains *Lactobacillus acidophilus*, a probiotic lactic acid bacterium that is beneficial to human health.

The *Lactobacillus acidophilus* bacteria use lactose for an energy source and reduce the amount of lactose present in milk. They also make the lactase enzyme which assists humans with lactose digestion in the small intestine.

Specialty milk beverages are available that are tailored to specific segments of the population. There are milk beverages with added plant sterols aimed at helping to improve cholesterol levels and others that are fortified with protein and calcium designed for adults. There are carbohydrate-reduced and vitamin fortified milk beverages for people watching their weight. Milk beverages targeted for teen athletes are protein fortified and fat-reduced. Milk beverages designed for children are calcium fortified, fat-reduced and flavoured. The flavoured milks compete with soft drinks for children's attention and come in a wide range of flavours from the traditional chocolate and strawberry to milks flavoured like their favourite candy bar or ice cream.

Yogurt Production

Yogurt Definitions

Yogurt is a fermented milk product that contains the characteristic bacterial cultures *Lactobacillus bulgaricus* and *Streptococcus thermophilus*. All yogurt must contain at least 8.25% solids not fat. Full fat yogurt must contain not less than 3.25% milk fat, lowfat yogurt not more than 2% milk fat, and nonfat yogurt less than 0.5% milk. The full legal definitions for yogurt, lowfat yogurt and nonfat yogurt are specified in the Standards of Identity listed in the U.S. Code of Federal Regulations (CFR), in sections 21 CFR 131.200, 21 CFR 131.203, and 21 CFR 131.206, respectively.

The two styles of yogurt commonly found in the grocery store are set type yogurt and swiss style yogurt. Set type yogurt is when the yogurt is packaged with the fruit on the bottom of the cup and the yogurt on top. Swiss style yogurt is when the fruit is blended into the yogurt prior to packaging.

Ingredients

The main ingredient in yogurt is milk. The type of milk used depends on the type of yogurt – whole milk for full fat yogurt, lowfat milk for lowfat yogurt, and skim milk for nonfat yogurt. Other dairy ingredients are allowed in yogurt to adjust the composition, such as cream to adjust the fat content, and nonfat dry milk to adjust the solids content. The solids content of yogurt is often adjusted above the 8.25% minimum to provide a better body and texture to the finished yogurt. The CFR contains a list of the permissible dairy ingredients for yogurt.

Stabilizers may also be used in yogurt to improve the body and texture by increasing firmness, preventing separation of the whey (syneresis), and helping to keep the fruit uniformly mixed in the yogurt. Stabilizers used in yogurt are alginates (carageenan), gelatins, gums (locust bean, guar), pectins, and starch.

Sweeteners, flavours and fruit preparations are used in yogurt to provide variety to the consumer. A list of permissible sweeteners for yogurt is found in the CFR.

Bacterial Cultures

The main (starter) cultures in yogurt are *Lactobacillus bulgaricus* and *Streptococcus thermophilus*. The function of the starter cultures is to ferment lactose (milk sugar) to produce lactic acid. The increase in lactic acid decreases pH and causes the milk to clot, or form the soft gel that is characteristic of yogurt. The fermentation of lactose also produces the flavour compounds that are characteristic of yogurt. *Lactobacillus bulgaricus* and *Streptococcus thermophilus* are the only 2 cultures required by law (CFR) to be present in yogurt.

Other bacterial cultures, such as *Lactobacillus acidophilus*, *Lactobacillus subsp. casei*, and Bifido-bacteria may be added to yogurt as probiotic cultures. Probiotic cultures benefit human health by improving lactose digestion, gastrointestinal function, and stimulating the immune system.

Cheese Production

Cheese Definitions

Cheese comes in many varieties. The variety determines the ingredients, processing, and characteristics of the cheese. The composition of many cheeses is defined by Standards of Identity in the U.S. Code of Federal Regulations (CFR).

Cheese can be made using pasteurized or raw milk. Cheese made from raw milk imparts different flavours and texture characteristics to the finished cheese. For some cheese varieties, raw milk is given a mild heat treatment (below pasteurization) prior to cheese making to destroy some of the spoilage organisms and provide better conditions for the cheese cultures. Cheese made from raw milk must be aged for at least 60 days, as defined in the CFR, section 7 CFR 58.439, to reduce the possibility of exposure to disease causing microorganisms (pathogens) that may be present in the milk. For some varieties cheese must be aged longer than 60 days.

Cheese can be broadly categorized as acid or rennet cheese, and natural or process cheeses. Acid cheeses are made by adding acid to the milk to cause the proteins to coagulate. Fresh cheeses, such as cream cheese or queso fresco, are made by direct acidification. Most types of cheese, such as cheddar or Swiss, use rennet (an enzyme) in addition to the starter cultures to coagulate the milk. The term "natural cheese" is an industry term referring to cheese that is made directly from milk. Process cheese is made using natural cheese plus other ingredients that are cooked together to change the textural and/or melting properties and increase shelf life.

Ingredients

The main ingredient in cheese is milk. Cheese is made using cow, goat, sheep, water buffalo or a blend of these milks.

The type of coagulant used depends on the type of cheese desired. For acid cheeses, an acid source such as acetic acid (the acid in vinegar) or gluconodelta-lactone (a mild food acid) is used. For rennet cheeses, calf rennet or, more commonly, a rennet produced through microbial bioprocessing is used. Calcium chloride is sometimes added to the cheese to improve the coagulation properties of the milk. Flavorings may be added depending on the cheese. Some common ingredients include herbs, spices, hot and sweet peppers, horseradish, and port wine.

Bacterial Cultures

Cultures for cheese making are called lactic acid bacteria (LAB) because their primary source of energy is the lactose in milk and their primary metabolic product is lactic acid. There is a wide variety of bacterial cultures available that provide distinct flavour and textural characteristics to cheeses.

Starter cultures are used early in the cheese making process to assist with coagulation by lowering the pH prior to rennet addition. The metabolism of the starter cultures contribute desirable flavour compounds, and help prevent the growth of spoilage organisms and pathogens. Typical starter bacteria include *Lactococcus lactis* subsp. *lactis* or *cremoris*, *Streptococcus salivarius* subsp. *thermophilus*, *Lactobacillus delbruckii* subsp. *bulgaricus*, and *Lactobacillus helveticus*.

Adjunct cultures are used to provide or enhance the characteristic flavours and textures of cheese. Common adjunct cultures added during manufacture include *Lactobacillus casei* and *Lactobacillus plantarum* for flavour in Cheddar cheese, or the use of *Propionibacterium freudenreichii* for eye formation in Swiss. Adjunct cultures can also be used as a smear for washing the outside of the formed cheese, such as the use of *Brevibacterium linens* of gruyere, brick and limburger cheeses.

Yeasts and molds are used in some cheeses to provide the characteristic colours and flavours of some cheese varieties. Torula yeast is used in the smear for the ripening of brick and limberger cheese. Examples of molds include *Penicillium camemberti* in camembert and brie, and *Penicillium roqueforti* in blue cheeses.

Ice Cream Production

Definitions

Ice cream is a frozen blend of a sweetened cream mixture and air, with added flavourings. A wide variety of ingredients are allowed in ice cream, but the minimum amounts of milk fat, milk solids (protein + lactose + minerals), and air are defined by Standards of Identity in the U.S. Code of Federal Regulations (CFR), section 21 CFR 135.110 for ice cream, 21 CFR 135.115 for goat's milk ice cream, and 21 CFR 135.140 for sherbet.

Ice cream must contain at least 10% milk fat, and at least 20% total milk solids, and may contain safe and suitable sweeteners, emulsifiers and stabilizers, and flavouring materials. The finished ice cream must weigh at least 4.5 pounds per gallon and there must be at least 1.6 pounds of total solids (fat + protein + lactose + minerals + added sugar) per gallon, thus limiting the maximum amount of air (called overrun) that can be incorporated into ice cream. There are well-defined labelling requirements for the types of flavours used (natural and/or artificial) and for the presence of egg yolks in the finished product (ice cream can be called custard or "French" if the content of egg yolks is at least 1.4%). Ice cream may also be labelled as reduced fat (25% less fat than the reference ice cream), light (50% less fat than the reference), lowfat (less than 3 g fat/serving), or nonfat (less than 0.5 g fat/serving).

Ice cream is sold as hard ice cream or soft serve. After the freezing process only a portion of the water is actually in a frozen state. Soft ice cream is served directly from the freezer where only a small amount of the water has been frozen. Hard ice cream is packaged from the freezer and then goes through a hardening process that freezes more of the water in the mix.

Ingredients

Milk fat provides creaminess and richness to ice cream and contributes to its melting characteristics. The minimum fat content is 10% and premium ice creams can contain as much as 16% milk fat. Sources of milk fat include milk, cream, and butter.

The total milk solids component of ice cream includes both the fat and other solids. The other milk solids consists of the protein and lactose in milk and ranges from 9 to 12% in ice cream. The nonfat solids play an important role in the body and texture of ice cream by stabilizing the air that is incorporated during the freezing process. Sources of nonfat solids include milk, cream, condensed milk, evaporated milk, dry milk, and whey.

Sweeteners are used to provide the characteristic sweetness of ice cream. Sweeteners also lower the freezing point of the mix to allow some water to reamin unfrozen at serving temperatures. A lower freezing point makes ice cream easier to scoop and eat, although the addition of too much sugar can make the product too soft. Sweeteners used include sugar (sucrose) and corn syrups.

Stabilizers are proteins or carbohydrates used in ice cream to add viscosity and control ice crystallization. Over time during frozen storage small ice crystals naturally migrate together and form larger ice crystals. Stabilizers help to keep the small crystals isolated and prevent the growth of large crystals, which causes ice cream to be coarse, icy and unpleasant to eat. Stabilizers used include alginates (carageenan), gums (locust bean, guar), and gelatins.

Emulsifiers are used to help keep the milk fat evenly dispersed in the ice cream during freezing and storage. A good distribution of fat helps stabilize the air incorporated into the ice cream and provide a smooth product. Emulsifiers used in ice cream include egg yolks and mono-and diglycerides.

A wide range of flavourings are used in ice cream. Flavourings include natural and artificial flavours, fruit, nuts, and bulky inclusions such as chocolate chunks and candies.

Dairy Accounting

To facilitate the recording and accounting of milk supplies each supplier should be given a code number. This code should have two elements:

1. A code for the producer and
2. A code for the dairy cooperative or peasants' association of which the producer is a member; a register of producers and their corresponding codes should be kept at the dairy centre.

The code is recorded in the milk record book, with the weight of milk received, and also on the sample bottle. The supplier should be given a copy of this record each week. He/she should be informed each day of the quality of the milk delivered.

Once the products to be made from the milk have been decided and the prices of the products determined, milk price can be calculated

Heat Treatments and Pasteurization

as follows, assuming that butter and cottage cheese are the chosen products:

1. Calculate the value of 1 kg of butterfat from the known price of butter, e.g. EB 10.

   Butter comprises 80% butterfat. Other constituents are regarded as having no commercial value. Therefore, the price of 1 kg of butterfat :

   = 10 × 100/80 = EB 12.5

2. Calculate the value of 1 litre of skim milk.

   Cottage cheese made from fermented skim milk has a value of EB 1.50/kg. Since there is about an 8-fold concentration of casein in the manufacture of cottage cheese from skim milk, assume an average yield of 1 kg of cottage cheese from 8 litres of skim milk. Therefore, each litre of skim milk has a value of 18 cents.

3. Calculation of the value of milk received: Assume the producer delivers 100 litres of milk containing 4% butterfat.

   1. Calculate the weight and value of butterfat received:

      The specific gravity of milk is 1.032 kg/litre. Therefore, the weight of milk received:

      = 100 litres × 1.032 = 103.2 kg

      Weight of fat received can be calculated by multiplying the weight of milk received by the fat content:

      = 103.2 × 0.04 = 4.128 kg

      Value of butterfat purchased from the producer is equal to the weight of butterfat received multiplied by the price per kg of butterfat:

      = 4.128 × EB12.5 = EB51.60

   2. Calculate the volume and value of the skim milk. While the actual recovery of skim milk may be greater, in commercial practice it is normally assumed that 80% of the whole milk is recovered as skim milk.

      In this case, we therefore recover 80 litres of skim milk with a value of 18 cents/litre.

      Value of skim milk = 80 × 0.18 = EB 14.40

   3. To obtain the total value of the milk received, add the values obtained in 3.1 and 3.2:

EB 51.60 for butterfat
EB 14.40 for skim milk
EB 66.00

Therefore, the average value of 1 litre of milk is 66 cents.

It is important to note that, since the butterfat is the most valuable commercial fraction, milk price will vary in proportion to butterfat content.

It is assumed that butterfat content can be estimated. In large dairy plants, milk price is based on the content of the major milk constituents. For small-scale milk processors, this is not normally feasible and payment should be based on fat content.

Production costs and depreciation are deducted proportionally from milk price. Other deductions may also be made when calculating the price paid to the producer for milk.

Milk Analysis

Milk analysis is carried out to determine:

- Freshness
- Adulteration
- Bacterial content, and
- Milk constituents for payment calculation.

Sampling

A representative sample is essential for accurate testing. Milk processors usually pay for milk or cream on the basis of butterfat analysis, and a single butterfat test may be used to determine the butterfat content of thousands of litres of milk or cream. Therefore, an accurate and representative sample must be obtained.

Milk must be mixed thoroughly prior to sampling and analysis to ensure a representative sample. If the volume of milk is small, e.g. from an individual cow, the milk may be poured from one bucket to another and a small sample of milk taken immediately. But if large volumes of milk are handled, the milk or cream must be mixed by stirring. However, it is very difficult to obtain a representative sample of milk or cream when a large volume is dumped into a large container. In such a case the milk must be stirred thoroughly and small samples taken from three or more places in the container. For best results, milk or cream must be sampled when it is at a temperature between

15 and 32°C. If the cream is too cool it will be thick and viscous and will be difficult to sample. Sour milk or cream, in which casein has coagulated, must be sampled frequently. Sampling sour milk follows the same procedure as for fresh milk. If the milk or cream has been standing for a long time and a deposit has formed on the surface and sides of the container, it should be warmed while agitating before a sample is removed. For certain analyses, milk samples can be preserved and stored to await analysis. Samples of milk or cream for butterfat analysis can be preserved using formalin, corrosive sublimate or potassium dichromate. For general analyses, formalin is preferred, because the other two increase the solids content of the milk, influencing total solids determination.

Estimation of Milk PH by Indicator

A rough estimate of pH may be obtained using paper strips impregnated with an indicator. Paper strips treated with bromocresol purple and bromothymol blue are sometimes used on creamery platforms as rejection tests for milk. Bromocresol purple indicator strips change from yellow to purple between pH 5.2 and 6.0, while bromothymol blue indicator papers change from straw yellow to blue-green between pH 6.0 and 6.9.

Electrometric Measurement of PH

Electrometric determination of pH depends on the potential difference set up between two electrodes when they are in contact with a test sample. A reference electrode whose potential is independent of the pH of the solution and an electrode whose potential is proportional to the hydronium ion concentration of the test sample are used. Saturated calomel electrodes are usually used as reference electrodes, and glass electrodes are used to measure pH. Instruments which measure the current produced by the difference in potential between the glass and calomel electrodes are called pH meters.

Preparation of the PH Meter

1. The pH meter should be kept in a dry atmosphere.
2. Before using a new glass electrode, or a glass electrode which has been stored for some time, soak the electrode in N/10 HCl for about 5 hours.
3. Care should be taken not to scratch glass electrodes against the sides of beakers or other hard surfaces during storage or testing.

4. The level of saturated potassium chloride in the calomel electrode should be checked before making pH measurements.
5. Crystals of potassium chloride should be present in the solution within the electrode.
6. The rubber stopper or cap on the filling arm of the calomel electrode should be removed before making a test.

Standardising and using the pH Meter

1. Rinse the electrodes with distilled water and wipe them gently with tissue or filter paper.
2. Set the temperature; use the control knob of the meter to set the temperature of the buffer used to standardise the meter.
3. Standardise the pH meter against a buffer solution of known pH. Use a buffer solution with a pH as close as possible to that of the test solution.
4. Turn the range selector to the pH range covering the pH of the buffer control until the pointer of the meter reads the pH of the buffer.
5. Set the range switch to zero.
6. Before measuring the pH of the test sample, rinse the electrodes with distilled water and dry them.
7. Set the temperature control knob to the temperature of the sample.
8. Place the test sample in position and allow the electrodes to dip into the solution.
9. Switch the range selector knob to the proper range and read the pH.
10. Rinse the electrodes after use and keep the electrode tips in distilled water between tests.

Always follow the manufacturer's instruction for the particular instrument.

Determination of Milk Acidity

The production of acid in milk is normally termed "souring" and the sour taste of such milk is due to lactic acid. The percentage of acid present in dairy products at any time is a rough indication of the age of the milk and the manner in which it has been handled.

Heat Treatments and Pasteurization 215

As mentioned earlier, fresh milk has an initial acidity due to its buffering capacity.

Apparatus
- White enamelled or porcelain cup
- Stirring rod
- A 10 ml or 17.6 ml pipette
- Burette
- Burette-stand.

Reagents
- One percent alcoholic solution of phenolphthalein
- N/10 or N/9 sodium hydroxide.

  I. Using N/10 sodium hydroxide.

Procedure
1. Fill the burette with N/10 NaOH and make sure there are no air bubbles trapped in the lower part.
2. Adjust the level of NaOH in the burette to the top mark – the lowest reading being at the upper end.
3. If milk, skim milk or buttermilk is to be tested, place 18 g in the cup using a 17.6 ml pipette. If cream is to be tested, use a 9 ml pipette (for cream weighing about 1 g/ml).
4. Add 3 to 5 drops of phenolphthalein to the sample in the cup.
5. Note the reading of the NaOH in the burette at the lowest point of the meniscus.
6. Allow the NaOH to flow slowly into the cup containing the sample and stir continuously. When a faint but definite pink colour persists, the end-point has been reached.
7. Take the reading of the burette at the lowest point of the meniscus. Subtract the first reading from the second to determine the number of millilitres of alkali (NaOH) required to neutralise the acid in the sample.

Calculation
Percent lactic acid = ml N/10 alkali × 0.0009 × 100/grams of sample II. Using N/9 sodium hydroxide: Milk, skim milk and buttermilk.

Apparatus

Same as for I.

Reagents
- 1.6% alcoholic solution of phenolphthalein.
- N/9 sodium hydroxide.

Procedure

1. Put 10 ml of milk in a porcelain dish.
2. Add 0.5 ml of 1.6% solution of phenolphthalein.

Titrate with N/9 sodium hydroxide and follow the same procedures as in I.

Calculation

Percent lactic acid = W/V

Where W = volume of N/9 NaOH required (ml) and

V = volume of milk taken for analysis (10 ml)

III. Using N/9 sodium hydroxide: Cream.

Procedure

1. Put 10 ml of cream in a porcelain dish.
2. Add 10 ml of water with the same pipette.
3. Add 0.5 ml of 1.6% phenolphthalein.
4. Titrate with N/9 NaOH.
5. Calculate as in II.

For determination of acidity of cream serum, the fat percentage of the cream should be known, and the calculation is as follows:

Acidity of serum = (acidity of cream × 100)/100 − % fat

Alcohol Test

The alcohol test, together with the acidity test, is used on fresh milk to indicate whether it will coagulate on processing. Milk that contains more than 0.21 % acid, or calcium and magnesium compounds in greater than normal amounts, will coagulate when alcohol is added.

Apparatus

- Ordinary 6-inch (15 cm) test tubes.
- Test-tube racks or blocks of wood with holes bored to fit the test tubes.

Reagents

The only reagent needed is a 75% alcohol solution. This is usually prepared from 95% alcohol by mixing with distilled water in the proportion of 79 parts of 95% alcohol to 21 parts of distilled water.

Procedure

1. Put equal volumes of milk and 75% alcohol in a test tube.
2. Invert the test tube several times with the thumb held tightly over the open end of the tube.
3. Examine the tube to determine whether the milk has coagulated: if it has, fine particles of curd will be visible.

Clot-on-boiling Test

Acidity decreases the heat stability of milk. The clot-on-boiling test is used to determine whether milk is suitable for processing, as it indicates whether milk is likely to coagulate during processing (usually pasteurisation). It is performed when milk is brought to the processing plant — if the milk fails the test it is rejected. The test measures the same characteristics as the alcohol test but is somewhat more lenient (0.22 to 0.24% acidity, as opposed to 0.21 % for the alcohol test). It has the advantage that no chemicals are needed. However, its disadvantage is that at high altitude milk (and all liquids) boils at lower temperature and therefore the test is even more lenient.

Apparatus

- One boiling water bath (a 600 ml beaker on a heater is adequate).
- Test tubes.
- Timer (a watch or clock is adequate).

Reagents

None

Procedure

1. Place about 5 ml of milk in a test tube (the exact amount is not critical), and place the test tube in boiling water for 5 minutes.
2. Carefully remove the test tube and examine for precipitate. The milk is failed if any curd forms.

Butterfat Determination

The main tests used to determine the fat content of milk and milk products are the Gerber and Babcock tests. Automated methods for testing milk are now used in central laboratories and at large processing centres.

The Gerber Test

The procedures outlined below are used to determine the butterfat content of milk, skim milk, buttermilk, cream and whey.

Milk

Apparatus

The apparatus required for butterfat content analysis comprises:
1. Gerber butyrometer calibrated to read 0–8% or 0–5% and graduated at 0.1 % intervals.
2. Butyrometer stoppers.
3. Milk pipette — volume to match the butyrometer in use.
4. 10 ml double-bulb pipette* for pipetting sulphuric acid.
5. 1 ml bulb pipette* for pipetting amyl alcohol.
6. Thermometer to read 1–100°C
7. Water bath.
8. Gerber centrifuge.

*Alternatively, automatic dispensers can be used for delivering 10 ml of sulphuric acid and 1 ml of amyl alcohol.

Reagents

- 1.825 specific gravity sulphuric acid
- Amyl alcohol.

Procedure

1. Mix the milk sample (temperature about 20°C) thoroughly, taking care to minimise incorporation of air. Allow the sample to stand for a few minutes to discharge any air bubbles. Mix gently again before pipetting.
2. Pipette or dispense 10 ml of sulphuric acid into the butyrometer.
3. Pipette the required volume of milk into the butyrometer. Care must be taken to avoid charring of the milk, by ensuring that

Heat Treatments and Pasteurization

the milk flows gently down the inside of the butyrometer. It then rests on top of the acid.
4. Pipette or dispense 1 ml of amyl alcohol.
5. Clean the neck of the butyrometer with a tissue or dry cloth.
6. Stopper the butyrometer tightly using a clean, dry stopper. Shake and invert the butyrometer several times until all the milk has been absorbed by the acid.
7. Then place the butyrometer in a water bath at 65°C for 5 minutes.
8. Centrifuge for 4 minutes at 1100 rpm.
9. Return the butyrometer to the water bath for 5 minutes, ensuring that the water level is high enough to heat the fat column.
10. Read the fat percentage. If necessary, the fat column can be adjusted by regulating the position of the stopper.

Hazards
- Sulphuric acid is toxic, highly corrosive and will cause severe burning if it comes in contact with the skin or eyes.
- When mixing the butyrometer contents, considerable heat is generated.
- If the stopper is slightly loose, leakage may occur during mixing, centrifuging or holding in the water bath.

Precautions
- Wear protective eye goggles
- Avoid all spillage and dropping of sulphuric acid from acid dispensers.
- When mixing, hold the butyrometer stopper firmly to ensure that it cannot slip. Use a cloth or glove to protect the hands when mixing.
- Do not point the butyrometer at anyone when mixing.

Skim milk, buttermilk and whey

Apparatus
Standard Gerber butyrometers designed for testing skim milk. The rest of the apparatus is the same as that used for whole milk.

Reagents: The same reagents are required as for whole milk.

Procedure

The procedure is the same as for whole milk up to and including the first centrifuging. The butyrometers are then placed in the water bath at 65°C, stoppers down, for 1 to 2 minutes and again centrifuged for 4 to 5 minutes.

Then they are placed in the water bath for 2 to 3 minutes and read. A check reading is made after they are placed in the water bath for 2 to 3 minutes. The readings obtained must be corrected as follows:

Percentage read on the:

butyrometer	Correction
<0.10%	Add 0.05%
0.10 to 0.25%	Add 0.02%
>0.25%	No correction required

Cream

Apparatus

The apparatus required for whole milk, except for the butyrometers and the 11 ml pipette, is supplemented by certain additional items for testing cream. The test bottles are standard Gerber cream butyrometers.

Other items include a balance for weighing to 0.001 or 0.005 g; a stand to support the butyrometers on the balance or a stopper weighing funnel, and a wash bottle containing warm (30–40°C) distilled water.

Reagents

The same as for whole milk.

Procedure

1. Mix the sample thoroughly, though cautiously, to avoid frothing. If the sample is very thick, it should be warmed to between 37.8° and 50°C to facilitate mixing.
2. Weigh 5 g of cream into the butyrometer.
3. Add about 6 ml of warm distilled water from the wash bottle.
4. Add 10 ml of sulphuric acid and I ml of amyl alcohol.

The remaining procedures are the same as for whole milk.

Cheese

Fat determination in cheese is carried out in a similar manner to that for milk.

Apparatus

Gerber cheese butyrometer stamped "3 g cheese". Other apparatus same as for Gerber milk fat analysis.

Reagents

- Distilled water
- Sulphuric acid
- Amyl alcohol.

Procedure

1. Weigh out 3 ± 0.01 g of cheese on a counter-balanced piece of grease-proof paper.
2. Dispense 10 ml sulphuric acid into the butyrometer. Add 3 ml of water carefully so that it rests on the acid.
3. Wrap the 3 g of cheese in the grease-proof paper to form a cylinder that fits into the butyrometer.
4. Add a further 4 to 5 ml of water.
5. Add 1 ml of amyl alcohol.
6. Stopper the butyrometer securely and shake to dissolve the cheese. (It may be difficult to dissolve the cheese. If difficulty is experienced, place the butyrometer in the heated water bath and remove periodically for mixing until the cheese is fully dissolved.) Cheese butyrometers are centrifuged and read as for milk and cream.

Determination of Milk Specific Gravity

Specific gravity is the relation between the mass of a given volume of any substance and that of an equal volume of water at the same temperature.

Since 1 ml of water at 4°C weighs 1 g, the mass of any material expressed in g/ml and its specific gravity (both at 4°C) will have the same numerical value. The specific gravity of milk averages 1.032, i.e. at 4°C 1 ml of milk weighs 1.032 g.

Since the mass of a given volume of water at a given temperature is known, the volume of a given mass, or the mass of a given volume of milk, cream, skim milk etc can be calculated from its specific gravity. For example, one litre of water at 4°C has a mass of 1 kg, and since the average specific gravity of milk is 1.032, one litre of average milk will have a mass of 1.032 kg.

Apparatus

- Lactometer – this is a hydrometer (a device for measuring specific gravity) adapted to the normal range of the specific gravity of milk. It is usually calibrated to read in lactometer degrees (L) rather than specific gravity *per se*. The relationship between the two is:

  (L/1000) + 1 = specific gravity (sp. gr.)

  Thus, if L = 31, specific gravity = 1.031.

- A tall, wide, glass or plastic cylinder.
- A thermometer – the lactometer may have a thermometer incorporated.

Procedure

1. Heat the sample of milk to 40°C and hold for 5 minutes. This is to get all the fat into a liquid state since crystalline fat has a very different density to liquid fat, and fat crystallises or melts slowly. After 5 minutes, cool the milk to 20°C.
2. Mix the milk sample thoroughly but gently. Do not shake vigorously or air bubbles will be incorporated and will affect the result.
3. Place the milk in the cylinder. Fill sufficiently that the milk will overflow when the lactometer is inserted.
4. Holding the lactometer by the tip, lower it gently into the milk. Do not let go until it is almost in equilibrium.
5. Allow the lactometer to float freely until it reaches equilibrium. Then read the lactometer at the top of the meniscus. Immediately, read the temperature of the milk. This should be 20°C. If the temperature of the milk is between 17 and 24°C, the following correction factors are used to determine L:

Temp. (°C)	17	18	19	20	21	22	23	24
Correction	−0.7	−0.5	−0.3	—	+0.3	+0.5	0.8	1.1

e.g. The lactometer reading is 30.5 and the temperature is 23°C.
Corrected lactometer = Lc = 30.5 + 0.8 = 31.3

Calculations

All calculations always use Lc, the corrected lactometer reading. To calculate the specific gravity, divide the corrected lactometer reading by 1000 and add 1.

In our example: Sp. gr. = (31.3)/ 1000 + 1 = 1.0313

Determination of total solids (TS) and solids-not-fat (SNF) in milk

The total solids content of milk is the total amount of material dispersed in the aqueous phase, i.e.

$$SNF = TS - \% \text{ fat.}$$

The only accurate way to determine TS is by evaporating the water from an accurately weighed sample. However, TS can be estimated from the corrected lactometer reading. The results are not likely to be very accurate because specific gravity is due to water, material less dense than water (fat) and material more dense than water (SNF). Therefore, milk with high fat and SNF contents could have the same specific gravity as milk with low fat and low SNF contents.

$$TS = (Lc)/ 4 + (1.22 \times \text{fat \%}) + 0.72$$
$$SNF = TS - \text{fat \%}$$
$$Or = Lc / 4 + (0.22 \times \text{fat\%}) + 0.72$$

It should be noted that the relationship between Lc and TS varies from country to country depending on milk composition. The above formulae are called the Richmond formulae and were calculated for Great Britain.

Determination of moisture content of butter

Apparatus

- Aluminium, platinum, nickel or porcelain cup, flat bottomed, about 3 cm in diameter, and not less than 2.5 cm deep, with a spout.
- A glass stirring rod with widened flat end.
- A spoon or steel blade.
- A butter trier.
- Alcohol lamp or other means of heating the sample
- Accurate moisture balance.
- Iron tripod.
- Asbestos-centre wire gauze.

Procedure

1. Weigh 10 g of butter into the cup. Heat the butter over a low flame until it ceases foaming and a light-brown colour appears. When heating the sample, place the container on the asbestos-centre wire gauze on a tripod. This distributes the heat evenly across the bottom of the cup.
2. After the moisture is driven from the butter, allow the sample to cool and reweigh.

Calculations

Percentage moisture content of the butter is calculated as:

Moisture % = (Original weight − final weight)/ Original weight × 100

Factors Affecting Milk Composition

Farmers are paid for market milk by volume, provided the milk meets minimum standards of composition-not less than 3.2 per cent fat and 11.75 per cent total solids, on a weight to weight basis. There is no minimum requirement for levels of solids-not-fat or protein. Manufacturing milk is bought on its yield of fat and protein.

Milk Composition

Milk from Friesian-Holstein cattle typically contains 87.5 per cent water and 12.5 per cent total solids. The ranges in composition of milk solids are: fat, 3.2 to 4.6 per cent; protein, 2.8 to 3.5 per cent; lactose, 4.2 to 4.8 per cent and minerals 0.6 to 0.8 per cent. The composition of milk is influenced by non-nutritional and nutritional factors.

Non-nutritional Factors Affecting Milk Composition

Breeding

Breeding is of considerable importance, since fat and protein levels in the milk are heritable characteristics.

Gains in milk composition made from breeding are permanent and accumulate from year to year. Benefits of sire and cow selection, and of mating decisions made today, will continue to be realized in all future descendants of the herd. In this respect, selection is a very productive means of improving milk composition.

The use of breeding to improve milk composition must be clearly understood, since selection to improve one production trait may lead to a decline in another. Selection based on milk yield will result in an increase in milk, fat and protein yields, but will reduce fat and protein percentages.

Table 1: Predicted responses to selection.

The more +, the greater the response over time to selection criteria.

Select for:	Predicted response in-				
	Milk	Fat %	Fat yield	Protein %	Protein yield
Milk	+++	-	+++	-	+++
Fat %	-	++	+	++	0
Fat yield	++	+	+++	-	++
Protein %	-	+	-	++	0
Protein yield	++	0	+++	+	+++
Fat and protein yield	+++	+	+++	+	+++

Similarly, selection of sires on protein or fat percentage only will result in a reduction in milk yield with minimal improvement in protein or fat yield. Because manufacturing milk is paid for on the yield of solids, you should select sires on the basis of fat PLUS protein yield. This will result in an increase in milk yield, fat percentage and yield, and protein percentage and yield. Since the stage of lactation affects the percentage and yields of protein and fat, you need detailed herd test data to select cows for breeding or culling purposes.

Stage of Lactation

The composition of milk varies with the stage of lactation. Cows that calve in good condition produce milk with a high fat and protein content during early lactation. The percentages of both fat and protein decline during the first six to eight weeks of lactation, then progressively rise after the cow becomes pregnant to reach their highest levels in late lactation.

Age

Although fat and protein contents decrease with increasing age, these changes are small. Since the age structure of a herd is not readily changed, the age composition of the herd is unlikely to contribute significantly to herd variation in milk composition.

Seasonal Conditions

Environmental factors that affect feed intake can be associated with pronounced variations in milk yield and composition.

Temperatures consistently above 30C will reduce milk yield as well as the percentage of milk protein, because of a reduction in energy intake. Cows in early to mid-lactation and receiving little or no supplementation (that is, relying on high pasture intakes) will be affected the most by heat stress.

Mastitis

Clinical and subclinical mastitis decrease milk yield and so reduce fat and protein yields.

Nutritional Factors

Level of Feeding

Precalving: Increased feed intake in late pregnancy increases milk yield and the yields of fat and protein. Research has shown that for each 30 kg increase in liveweight at calving, milk yield increases by 122 kg, fat yield by 8 kg and protein yield by 4 kg during the first 20 weeks of lactation. The effects of condition score at calving on fat and protein percentage are small.

Post-calving: The effect of feeding level on fat and protein percentage is variable. This is because the stage of lactation influences the effect of feed intake on milk composition If feed intake is increased during early lactation, milk yield will increase with consequent increases in fat and protein yields. As intake increases, the percentage of milk fat will decline, but protein percentage will increase slightly. Protein production in well fed herds is rarely below 3.2 per cent, but in poorly fed herds it can fall to 2.8 per cent.

Diet Quality

Pasture species influence milk yield and composition, as shown in table below. The use of species associated with improved pasture quality results in increased milk, fat and protein yields.

Table 2: Effect of pasture species on milk production, as shown in two trials

	Trial 1		Trial 2	
	Kikuyu	Ryegrass	Ryegrass	Clover
Milk yield (L/day)	13.4	19.4	16.5	18.9
Milk Fat (%)	3.7	3.5	3.7	3.5
Milk Fat kg/day	0.51	0.70	0.59	0.69
Protein %	2.9	3.2	3.0	3.2
Protein kg/day	0.40	0.64	0.51	0.62

The increase in fat yield is caused by an increase in milk yield only, since the percentage of milk fat actually declines. increases in both milk yield and percentage protein cause the increase in protein yield. Species differences are largely caused by inherent differences in intake. However, with ryegrass and clovers, differences still occur when they are fed at the same level.

Concentrates

Providing supplementary feed in the form of cereal grain usually results in increased milk, fat and protein yields. An increase in milk yield causes the increase in fat yield, since the percentage of milk fat often declines.

Increases in both milk yield and protein percentage cause the increase in protein yield.

Feeding lupins also results in increases in milk fat and protein yields. Unlike the cereal grains, lupins do not reduce the fat percentage when they are fed as a supplement to cattle.

The cow generally uses protein supplements as a source of energy rather than a supply of protein to the udder. Providing there is sufficient protein in the total diet, feeding protein supplements will result in a similar increase in protein percentage as feeding a similar amount of energy from cereal grain.

If the protein content of the total diet is low, feeding protein supplements increases the energy content of the total diet by increasing the digestibility of the total diet. As a result, the protein percentage of the milk is increased.

Milk production and milk protein content will increase when an energy deficiency is corrected.

When concentrates are fed, the degree of processing can affect the fat percentage of the milk. Grains need only be cracked to allow sufficient digestion. Over-processing can reduce the fat percentage of the milk.

Fibre

If the milk fat percentage has dropped, but the protein percentage has remained constant, more fibre is needed in the total diet. This is best provided by feeding hay. However, cattle need only small quantities of hay (2 to 3 kg/cow/day)when they are grazing good quality pasture.

Summary

To increase fat and protein yields by feeding, increase the energy intake of the cow by:

- greater pasture intake, by increasing pasture availability (for example, by using irrigation or more nitrogenous fertilizers);
- greater pasture intake by improving the pasture quality, by species selection (ryegrass/clover) or improved grazing management (short compared with rank pasture); or
- by feeding supplements.

Cow's Genetic Predisposition Affects Composition of Her Milk

The genetic predisposition of cows has an effect on the fat and protein content of their milk. Researchers at Wageningen University have spent the past few years examining the scope and significance of genetic variation between cows for the differences in quality characteristics of milk. They have discovered a number of genes that contribute to this genetic variation.

The research was carried out as part of the large-scale Milk Genomics project that Wageningen University launched in 2004 in association with the cattle breeding and dairy sector. On the basis of this knowledge, it is possible to devise an innovative breeding programme for cows and bulls to increase the proportion of unsaturated fatty acids in the milk and to improve cheese production.

Researchers found enormous variation in the composition of the milk fat in cows' milk. A significant proportion of these differences can be put down to genetic predisposition. DNA was analysed to find out which genes contribute to genetic differences between the animals. Researchers in Wageningen demonstrated that a mutation in a gene with a large influence on the amount of fat in milk, also affects the composition of that milk fat. Moreover, the Wageningen researchers were able to make use of available cattle genome data in their research; late last week more than 300 scientists published on this subject in the journal *Science*.

The information on the cattle genome was used to identify new genes that affect the quality characteristics of milk. They identified six areas on the genome, where genes contributing to the genetic variation in milk fat composition are found. According to the researchers, these findings provide an opportunity to devise an innovative breeding programme that exploits the natural variation within the dairy cattle population to make a targeted selection of cattle that produce milk with a modified fat composition. They predicted that the proportion of unsaturated fatty acids in milk could increase by ten percent in ten years by selection of bulls, in addition to the influence that animal nutrition could have on this proportion.

Proteins

The researchers at Wageningen University also discovered substantial variation in the composition of milk proteins, which mainly comprise caseins. Here too, the differences can largely be put down to genetic variation.

DNA analysis revealed three areas on the cow genome affecting the protein composition. Unlike the composition of milk fat, the effect of feed on the differences in the composition of milk protein is relatively small. A higher proportion of caseins results in increased cheese production, which at a rough estimate represents an extra 25 million Euros for the Dutch dairy sector.

Five years ago, researchers at Wageningen University together with the NZO (Dutch Dairy Association) and CRV, an organisation that focuses on cattle improvement, set up the Milk Genomics initiative with the aim of identifying the genes that contribute to natural variation in the quality aspects of milk, and more specifically in composition of fat and proteins. The database created for the genetic research is unique in the world in terms of the scope and number of data. It consists of data on around 2,000 cows from 400 farms.

Milk Composition

Today, not only the nutritional value of milk but also other physiological properties of milk components have attracted interest. Bovine milk contains approximately 87% water, 4.6% lactose, 3.4% protein, 4.2% fat, 0.8% minerals and 0.1% vitamins. The composition of milk continuously undergoes changes depending on e.g. breeding, feeding strategies, management of the cow, lactation stage and season.

Milk Fat

The lipids in bovine milk are mainly present in globules as an oil-in-water emulsion. These fat droplets are formed by the endoplasmic reticulum in the epithelial cells in the alveoli and coated with a surface material of proteins and polar lipids.

When secreted, they are enveloped with the plasma membrane of the cell. Membrane-associated materials can comprise 2–6% of the globule mass. The composition and structure of the milk fat globule membrane (MFGM) is not known in detail but it is mainly composed of polar lipids and membrane-bound and associated proteins.

The lipid fraction comprising approximately 30% of the membrane material consists of lipids such as phospholipids (25%), cerebrosides (3%) and cholesterol (2%). The remaining 70% of the membrane material are proteins, many of them being enzymes.

The milk fat consists mainly of triglycerides, approximately 98%, while other milk lipids are diacylglycerol (about 2% of the lipid fraction), cholesterol (less than 0.5%), phospholipids (about 1%) and free fatty acids (FFA) (about 0.1). In addition, there are trace amounts of ether lipids, hydrocarbons, fat-soluble vitamins, flavour compounds and compounds introduced by the feed. The size of the milk fat globule (MFG) increases with increasing fat content in the milk probably because of a limitation in production of MFGM.

The number of MFG in milk is approximately 10^{10} per mL with a total area of 700 cm^2 per mL of milk. The size of the MFG has crucial influence on the stability and technological properties of milk. Milk lipid globules are resistant to pancreatic lipolysis in the small intestine unless they are first exposed to gastric lipolysis.

Origin of Milk Fatty Acids

The milk fatty acids are derived almost equally from two sources, the feed and the microbial activity in the rumen of the cow. The fatty acid synthesising system in the mammary gland of the cow produces fatty acids with even number of carbons of 4–16 carbons in length and accounts for approximately 60 and 45% of the fatty acids on a molar and weight basis, respectively.

This *de novo* synthesis in the mammary gland is of the 4:0–14:0 acids together with about half of the 16:0 from acetate and β-hydroxybutyrate. Acetate and butyric acid are generated in the rumen by fermentation of feed components. The butyric acid is converted to β-hydroxybutyrate during absorption through the rumen epithelium. Bovine fat contains certain fatty acids with odd number of carbons, such as pentadecanoic acid (15:0) and heptadecanoic acid (17:0). These two fatty acids are synthesised by the bacterial flora in the rumen. The remaining 16:0 and the long-chain fatty acids originate from dietary lipids and from lipolysis of adipose tissue triacylglycerols. Medium-and long-chain fatty acids, but mainly 18:0, may be desaturated in the mammary gland to form the corresponding monosaturated acids.

Fatty acids are not randomly esterified at the three positions of the triacylglycerol molecule. The short-chain acids butyric (4:0) and caproic (6:0) are esterified almost entirely at sn-3. Medium-chain fatty acids (8:0–14:0) as well as 16:0 are preferentially esterified at positions sn-1 and sn-2. Stearic acid (18:0) is selectively placed at position sn-1, whereas oleic acid (18:1) shows preference for positions sn-1 and sn-3.

When consumed by humans, milk triacylglycerols are lipolysed by lingual lipases in the mouth and by both lingual and gastric lipase in the stomach. The lipases preferentially hydrolyse sn-3 position fatty acids, and therefore selectively releases the shorter acids. The result is that 4:0–10:0 pass through the stomach wall in decreasing quantities as the molecular weight increases, enter the portal vein, and are transported to the liver where they are oxidised. About 25–40% of the triacylglycerols are digested in the stomach.

Fatty Acid Composition

Milk fat triacylglycerols are synthesised from more than 400 different fatty acids, which makes milk fat the most complex of all natural fats. Nearly all of these acids are present in trace quantities and only about 15 acids at the 1% level or higher.

Many factors are associated with the variations in the amount and fatty acid composition of bovine milk lipids (15, 16). They may be of animal origin, i.e. related to genetics (breed and selection), stage of lactation, mastitis and ruminal fermentation, or they may be feed-related factors, i.e. related to fibre and energy intake, dietary fats, and seasonal and regional effects. The gross composition of milk fat in Swedish dairy milk 2001 was 69.4% saturated fatty acids and 30.6% unsaturated fatty acids.

The content of saturated fatty acids is lowest in the summer when the cows are grazing, and highest in the winter due to indoor feeding. The content of the unsaturated fatty acids shows the opposite pattern with the highest amount in the summer.

The saturated fatty acids present in milk accounts for approximately 70% by weight (5). The most important fatty acid from a quantitative viewpoint is palmitic acid (16:0), which accounts for approximately 30% by weight of the total fatty acids. Myristic acid (14:0) and stearic acid (18:0) make up 11 and 12% by weight, respectively. Of the saturated fatty acids, 10.9% are short-chain fatty acids (C4:0–C10:0). The amounts of butyric acid (4:0) and caproic acid (6:0) on a yearly average are 4.4 and 2.4% by weight of the total fatty acids, respectively, in Swedish dairy milk. These amounts are higher when their proportions are expressed as molar percentages, approximately 10 and 5%, respectively.

Approximately 25% of the fatty acids in milk are mono-unsaturated with oleic acid (18:1) accounting for 23.8% by weight of the total fatty acids in Swedish dairy milk.

Poly-unsaturated fatty acids constitute about 2.3% by weight of the total fatty acids and the main poly-unsaturated fatty acids are linoleic acid (18:2) and α-linolenic acid (18:3) accounting for 1.6 and 0.7% by weight of the total fatty acids. The ratio between omega-6 and omega-3 fatty acids in Swedish milk fat was 2.3:1 in 2001. Milk and meat from ruminants can be an important source of omega-3 fatty acids in the human diet, as is the case in France, where animal products account for about 40% of the intake.

Approximately 2.7% of the fatty acids in milk are trans fatty acids with one or more trans-double bonds (18). The main trans 18:1 isomer is vaccenic acid (VA), (18:1, 11t), but trans double bounds in position 4–16 is also observed in low concentrations in milk fat. VA constitutes approximately 2.7% of the total fatty acid content and varies with season.

Milk fat contains also conjugated linoleic acid (CLA), with many different isomers including rumenic acid (RA) (cis-9, trans-11 CLA) which predominates (75–90% of total CLA). The majority of RA in the milk fat is synthesised endogenously, in the mammary gland through the action of mammary Ä-desaturase on VA. Thus both VA and RA are present in milk and dairy products, generally in the ration of about 1:3.

VA has a double role in metabolism because it is both a trans fatty acid and a precursor for 9c, 11t-CLA. A small amount of the CLA originates from biohydrogenation of unsaturated fatty acids by rumen bacteria. Animal studies and new human data have confirmed the bioconversion of VA into CLA (21, 22). Bovine milk, milk products and bovine meat are the main dietary sources of the RA (23). Milk content of 9c, 11t-CLA varies considerably, but in Swedish milk fat it constitutes about 0.4% of the fat fraction.

What is Milk?

Dairy herds consist of cows which produce large volumes of milk. All cows are female; males are called bulls. The most common dairy breed is the Holstein, the black and white cows often seen in pastures. Cows are mammals and like all mammals produce milk for their young. This is the milk we get from cows.

Where is Milk Produced in BC?

Most of the dairy herds are in the Lower Mainland, southeastern Vancouver Island, and north Okanagan-Shuswap area. 70% of BC's milk production is in the Fraser Valley. A few dairy herds, about 20%, are located in the North Okanagan, East Kootenay and Bulkley Valley/Cariboo/Peace regions and 10% on South-east Vancouver Island.

How Much Milk do we Produce?

About 842 dairy farms produce an annual volume of over 570 million litres a year. The average herd size is 80 cows plus additional replacement calves and heifers. The average cow on test produces 29L of milk a day and is milked for 10 months a year, which equals 8839 litres of milk per year per cow. That's an average of 100 glasses of milk per day, every day of the year. This amount would fill up 53 bathtubs.

How is Milk Produced?

Before any cow produces milk, she must first become a mother. When a dairy cow reaches about 15 months in age she is bred, usually by artificial insemination. After about 9 months she has a calf and produces milk. After a cow has given birth, she can produce milk for the next 10 months. A cow that is being milked can eat up to 40kg of grass, forage, and hay a day and drink up to 170L of water a day, especially on hot days. That's over a bathtub full. A cow's diet is supplemented with feeds such as barley, wheat, soybean and canola meal. These are formulated and fed according to the energy, protein and other nutritional needs of the animal.

Milking machines are used to milk a cow. The cows go into a milking barn, their udders are cleaned and a rubber lined suction cup is attached to the teat. The suction cup simulates the suckling action of a calf nursing. The suction cup is attached to hoses and pipes which collect the milk in a holding tank. The milk is then quickly cooled. Cows are milked twice and sometimes three times a day, usually at the same times each day. All equipment used for milking is thoroughly cleaned and sanitized before and after each use.

Dairy farms are inspected and certified before it can produce milk. This includes: all milking equipment, milking procedures, milking parlour and barn—everywhere the cows go must be kept clean and well maintained.

Dairy farmers use computers to keep track of how much each cow eats, how much milk each cow produces and even to match a particular cow with a particular bull for breeding. They also use them for finding information (internet) and financial accounting.

What does Milk look Like When We Use It?

We drink fresh milk (whole, 2%, 1%, skim and chocolate) and use milk products such as cheese, yogurt, sour cream, whipping cream, cottage cheese, evaporated milk, sweetened condensed milk and skim milk powder.

Among the cheeses BC produces are cheddar, mozzarella, Parmesan, colby, gouda, farmer, edam, monterey jack, feta, quark, cottage cheese and ricotta.

Milk is 89% water and 11% solids. The nutrients, such as calcium, riboflavin, vitamin A and protein are in the solids. Milk, cheese, and yogurt are an easy way for most people to get the amount of dietary calcium recommended by Health Canada.

What Happens after the Milk Leaves the Farm?

Milk is picked up from the farm by a tanker truck, that is certified before it can carry milk, every second day.

The licensed driver takes a sample from each farm to ensure the milk meets quality and safety standards. Before the milk can be unloaded it is tested for antibiotic residues. If residues are found, the entire shipment is destroyed and the farmer responsible receives a heavy fine (thousands of dollars) and also pays for the entire truckload of milk.

To ensure the safety of milk, it is pasteurized. This is the process of heating milk quickly to 72°C and cooling it very rapidly to 4°C. This kills any harmful bacteria that may find its way into milk. Pasteurizing milk helps keep milk fresh longer by destroying spoilage organisms. The milk is also tested by a certified laboratory for temperature, acidity and flavour, before it is accepted. Bacteria, water contamination and somatic cell counts are tests that are also done regularly. The presence of somatic cell counts is an indicator of animal health and quality. Other tests are carried out from time to time to ensure purity of the product.

In days gone by before homogenization, the cream always rose to the top. Today, most milk is homogenized. Homogenization ensures that the cream is thoroughly mixed throughout the product so that it does not separate out. This process doesn't alter any of the nutrients found in milk.

Throughout the entire process from the time the cow is milked until the milk is packaged, milk is never touched by human hands. Milk is natural-nothing is added except vitamin A and D which is required by law. Milk remains one of the purest and safest foods available.

The dairy (processing plant) is also inspected regularly for cleanliness, handling procedures and equipment standards. All milk contact equipment is cleansed and sanitized on a daily basis-failure to do so would result in bacterial spoilage before the code date. Every dairy and their employees who work in the processing area must be licensed.

Milk is packaged within days, usually within 24 hours, of arriving at a dairy plant. Packaged dairy products are also regularly tested by a certified laboratory for composition to ensure the product contains what it claims.

This is also the final check point to ensure the product meets the standards established of bacteria, coliforms, yeasts, moulds and other potential contaminants. Dairy products at retail outlets are subject to random sampling as a further check of their safety, quality and composition.

The majority of milk produced in BC is sold as fluid milk while the rest is manufactured into semi-fluid and solid products such as cheese, ice cream, yogurt and cottage cheese.

What Producing Challenges do Dairy Producers Face?

Dairy farms are close to being self-sustaining units. Farmers grow two-thirds of the food a cow eats and recycle the cow's manure back to the fields where the feed is grown. Manure is very useful to farmers because it adds nutrients and organic matter which help to sustain and build the quality of the soil.

Further challenges facing today's dairy producers include:

Meeting environmental requirements.

Surviving a market that is increasingly competitive on a global scale.

Increasing input costs for such things as feed (grain), equipment and labour, with decreasing revenue.

Dealing with increasing competition for land use (e.g., urban push, increasing land values, etc.).

Who's Involved in Getting the Milk From the Farm To the Table?
- Dairy farm owner, manager and staff (milkers, herdsmen, field personnel)
- Breed associations
- Artificial insemination technicians
- Dairy herd improvement advisors
- Veterinarians
- Milking equipment, farm equipment, building and facility suppliers
- Feed producers and nutritionists
- Dairy processor field representatives
- Government inspectors and advisors
- Government and university researchers
- Milk tank truck drivers
- Milk product deliverers
- Store employees.

7

Chemicals: Lead, Mercury, Cadmium and Other Metals

Metals play an important role in human biology, and trace amounts of some metals — manganese, for example — are essential to life. At higher concentrations, however, these same metals are toxic. In addition, some metals — lead, for example — do not occur naturally in the body, and their presence, usually as a result of occupational or pollution-related exposure, is detrimental to health. A number of potentially toxic metals have been reported in breast milk, including lead, mercury and cadmium. Unlike the persistent organic pollutants (POPs), metals do not accumulate in fat, and so do not usually achieve higher concentrations in breast milk than in blood. As a result, infants are likely to be exposed to higher levels before birth than during breastfeeding. Nonetheless, learning about metals in breast milk is important for two reasons: first, as a pathway of exposure, and second, as an indicator of likely prenatal exposures.

Many metals that have been reported in breast milk also contaminate drinking water and so can occur in infant formula at levels even higher than in breast milk. For example, an infant's exposure to cadmium from soy infant formula is about 20 times higher than the levels generally found in breast milk. Estimated cadmium intake from powdered formula has been estimated at six times higher than average levels in breast milk.

Health Effects of Metals

Lead has often been called the leading environmental health threat to children. It is toxic to the developing brain, and at high levels results in numerous poisoning symptoms. In addition, at the low doses common today in many countries, lead has subtle effects on neurological functions, including learning, memory and attention span. Because the infant brain is developing rapidly both before birth and for several years after birth, lead exposures during this critical period are particularly detrimental to the future intellectual potential of children.

Mercury occurs in a number of forms in the environment. The most hazardous for children is methyl mercury, although inorganic mercury is also a potential concern. Methyl mercury, like lead, is toxic to the developing brain. While high-dose exposures can result in a cerebral palsy-like syndrome, low dose exposures may cause subtle deficits in learning and memory. Other metals, such as cadmium, arsenic and manganese, have not been as thoroughly studied in breast milk. Arsenic is known to cause cancer in humans, and high levels of manganese can cause a syndrome that resembles Parkinson's disease. Cadmium is toxic to the male reproductive system, the kidneys, bone and the brain. All of these contaminants are more likely to affect bottle-fed infants because they are water contaminants and are often found at higher concentrations in infant formula as compared with breast milk.

Human Exposure

People can be exposed to metals in a number of ways, including at work in certain industries, from drinking contaminated water and eating contaminated food, or in hobbies that involve working with metals.

Lead exposure stems primarily from its use in gasoline, paint, water pipes and the lining of food cans. These uses have been banned in many countries, but still persist in many parts of the world. In addition, old, peeling paint and old water pipes can still cause exposures. Other common sources of lead include:
- painting or removing old paint;
- construction work;
- battery manufacturing or recycling;
- automobile repair;
- electronics work;
- ceramics and pottery glazed with lead;
- welding and soldering;
- firearm shooting and cleaning;

- jewellery making and repair;
- stained-glass-window making; and
- cosmetics, including certain hair dyes and kohl.

Most people are exposed to methyl mercury from fish, particularly such predator fish as swordfish, shark and tuna. Freshwater fish from contaminated lakes, rivers and estuaries can also bioaccumulate very high levels of methyl mercury, which are passed on to humans who eat the fish. Other sources of mercury include coal burning, incineration, chlorine manufacturing and mining, as well as some natural sources. Inorganic mercury exposure primarily comes from dental amalgam fillings. Exposure to cadmium often comes as a result of work or through hobbies, including metal plating, semiconductor manufacture, welding, soldering, ceramics and painting. In addition, it is a contaminant of drinking water, air and food, particularly shellfish. One other important source of cadmium is cigarette smoke; smokers typically have blood levels of cadmium approximately twice those of nonsmokers.

Breast Milk

In general, the metals found in breast milk are usually at lower levels than are found in maternal blood. Thus, breast milk is not the primary pathway of exposure for infants; prenatal trans-placental exposure is a much greater concern. That said, instances of high exposure through breast milk do occur, and are often important indicators of an infant's total exposure. One study found that longer duration of breastfeeding was associated with poorer infant growth in children whose mothers had higher levels of mercury in their bodies. Generally, infants fed formula made with tap water are at the highest risk from metals contaminating the water supply.

Bans and Restrictions

In recent years, some bans and restrictions regarding certain heavy metals have been implemented. For example, lead has been banned from use in gasoline, paint, can linings, or water pipes in more than 50 countries.

The past few years have brought a great deal of state and local legislation restricting the use of mercury in products. For example, several states have passed laws or regulations governing the sale and use of mercury, requiring recycling and imposing notification requirements on the substance and many products containing it.

There are many restrictions already in place relating to mercury containing thermometers, and many local communities have set up mercury fever thermometer exchange programs. Several states have adopted bans on offering for sale or use, or distributing for promotional purposes, mercury-added novelties. These are products intended mainly for personal or household enjoyment or for adornment, such as toys, games, ornaments, holiday decorations, apparel, jewellery, figurines and yard statues.

Cadmium has also been the subject of legislation. The European Union banned the use of cadmium in materials and components of vehicles put on the market after July 1, 2003, and in new electrical and electronic equipment after July 1, 2006.

Some countries have also had success reducing pollution from incinerators, power plants and factories, thereby reducing emissions of mercury and other metals.

Benchmarks and Exposure Limits

The U.S. EPA has set an action level for lead in water of 15 parts per billion (ppb) and an action level for inorganic mercury of 2 ppb. While the median level of lead in breast milk worldwide is only one-third as high as the U.S. drinking water limit, the most exposed populations have lead levels in breast milk that exceed this limit threefold. Average levels of mercury in breast milk are near the action level for water.

The World Health Organization (WHO) has set a daily permissible intake (DPI) level of 5 micrograms per kilogram per day (µg/kg/day) of lead for children, and the DPI for cadmium is 1 µg/kg/day (for an adult). The U.S. Agency for Toxic Substance and Disease Registry (ATSDR) has established a minimal risk level (MRL) for mercury of 2 µg/kg/day for inorganic mercury and 0.12 µg/kg/day for methyl mercury.

On average, breastfeeding infants are unlikely to exceed these levels. However, in a few polluted communities around the world, infant exposures do exceed these levels.

In the United States, the Health Resources and Services Administration has published a blood lead action level for breastfeeding women of 40 micrograms per deciliter (µg/dL/day)of blood or above.

Women with lead levels this high in their blood may be encouraged to choose alternatives to breastfeeding. Most women have blood lead levels far below 40 micrograms per deciliter. However, recent studies looking at children with blood lead levels of 10 micrograms per dL or less and pregnant women with blood lead levels less than 5 microg/dL have demonstrated significant effects on the children's memory and cognitive functioning. This suggests that the current lead standards may still be too high.

Breast Milk Monitoring Studies

Metals have been detected in breast milk around the world. Countries that have conducted studies detecting one or more of the three major metals of concern (cadmium, lead and mercury) include:

Austria	Hungary	Netherlands	Spain
Belgium	India	New Zealand	Sweden
Bulgaria	Iraq	Nigeria	Thailand
Canada	Italy	Pakistan	Turkey
Finland	Japan	Philippines	United States
Germany	Malaysia	Poland	United Kingdom
Guatemala	Mexico	Romania	Yugoslavia

Levels of lead, manganese and mercury vary widely in breast milk samples around the world, with very high levels detected in some places. Results of a WHO study on trace elements in breast milk are summarized below. These levels reflect both maternal-absorbed dose of these metals and infant exposure, and illustrate the large ranges of exposure across the population. Among women who eat a lot of fish, for example, levels of mercury in breast milk may exceed levels in unexposed women by 100 times.

Table 1: Range of Metals Detected in Breast Milk Around the World

Metal	*Median (ppb)*	*Range (ppb)*
Arsenic	0.3	0.1-0.8
Cadmium	0.1	0.1-3.8
Lead	5.0	0.0-41.1
Manganese	18.0	7-102.0
Mercury	2.7	0.64-257.1

Source: World Health Organization, 1993

ppb = parts per billion

Important Case Examples: Several specific examples of metals in the human body merit examination.

Lead

The elevated presence of lead in human blood samples has been an issue for decades, chiefly because of two products: lead-based gasoline and paint. Lead is ubiquitous in the environment as a result, and in many areas of the world, a significant level of lead turns up in breast milk (5 to 20 ppb). That said, lead does not concentrate in breast milk because it does not attach to fat; indeed, levels of lead are generally higher in a mother's blood than in her milk. Several studies have found higher blood lead levels in formula-fed infants than in breast-fed infants. This may be a result of contaminated formula cans or formula prepared using tap water with high lead levels. Lead levels in blood and breast milk correlate closely with areas where lead is still used in gasoline, with the highest levels in areas with heavy traffic. In addition, mothers in countries where lead is still used in gasoline, and mothers living near lead smelters, have higher levels of lead in their breast milk due to community contamination.

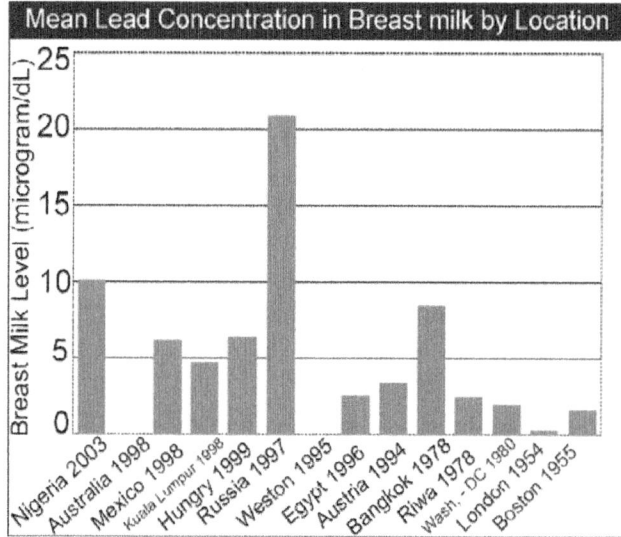

Much of the lead in breast milk does not come from the mothers' exposure during lactation. Instead, it comes from lead stored in the mothers' bones. Because lead mimics the beneficial mineral calcium, it is stored for decades along with calcium in the bones. During pregnancy and lactation, a woman's body extracts calcium from her own bones to provide calcium for her child's bone development. Calcium extraction from bone is greatest during lactation, and as a result, lead stored in the mother's bones also enters the blood and breast milk during pregnancy and lactation, posing an exposure risk to the fetus.

A study in April 2003 confirmed that ensuring adequate dietary calcium intake or taking a calcium supplement before pregnancy, during pregnancy and during the entire lactation period decreases the blood lead level in lactating women. Supplemental dietary calcium most likely decreases the amount of calcium and lead that comes out of the mother's bones. Therefore, women can significantly reduce their baby's exposure to lead by getting adequate dietary calcium or taking a calcium supplement during pregnancy and lactation.

Mercury

Breast milk levels of mercury are usually lower than levels of lead. Mercury does not accumulate in breast milk; in fact, the levels in the mother's blood are generally about three times higher than the levels in milk. Therefore, prenatal exposure is probably more important than lactation exposure to mercury, in most cases. Two major forms of mercury can enter breast milk. The most hazardous, methyl mercury, does not enter breast milk at high rates because it is attached to red blood cells. But what little does get into breast milk is easily absorbed in the intestine of a nursing infant.

The second form, inorganic mercury, enters breast milk easily but is not well absorbed in the infant's gastrointestinal system. One Swedish study found that the mercury in breast milk in the early months of breastfeeding was primarily inorganic mercury from dental amalgam fillings in the mother's mouth. However, after two months of lactation, mercury found in milk was primarily from methyl mercury associated with the mother's fish consumption, rather than dental amalgam fillings. In the past, mercury has been responsible for several mass poisonings — in Minamata, Japan, and in Iraq.

In both cases, food contaminated with methyl mercury led to illness and death. Some of those affected were breastfeeding children whose mothers had eaten the contaminated food. However, in both of these scenarios, the levels of mercury were extremely high. The average levels found in women's breast milk today are far lower than in those cases. In one recent study of mercury exposure, breast-fed infants tended to have higher residues of mercury detectable in their hair. The infants with higher hair mercury levels also had improved neurological development, including faster progression to sitting, creeping and standing. Because mercury is known to affect neurological development adversely, the faster development in infants with higher mercury levels was attributed to the benefits of breastfeeding. Thus any possible adverse effects of mercury in breast milk were overcome by the advantages of breastfeeding.

A recent study on laboratory rats lends support to the merits of breastfeeding with respect to mercury. The study looked at the effects of diet that contained a moderate dose of mercury — 5 parts per million — fed to pregnant rats before and during pregnancy and during lactation. The newborn rats on the day of birth actually had 1.5 times more mercury in their blood than their mother. However, during the lactation period the babies' blood mercury levels dropped dramatically.

Even though the breast milk contained some methyl-mercury, this study suggests that the exposure of the baby to mercury during breastfeeding is much lower than during fetal development. The diet used in this study simulated that of a high fish consuming-population. After weaning, the newborns were started on a mercury-containing diet and showed an increase in their blood mercury levels. However, a human study in the Faroe Islands, where people eat fish and whale meat, found that those children who were breastfed exclusively for six months ended up shorter and thinner than their peers at 18 months of age. When the researchers looked for an explanation for this finding, they discovered that the children who were exposed to the highest levels of mercury before birth and during breastfeeding were the ones with growth delays.

A number of studies have examined the protective effects of Vitamin E on mercury, finding that vitamin E may decrease the toxic effects of mercury. In addition, the trace element selenium may enhance the ability to clear mercury from the body's cells. This data further supports the role of supplemental multivitamins during pregnancy.

Cadmium

Cadmium levels in breast milk are significantly associated with cigarette smoking. One German study showed a direct relationship between the number of cigarettes a mother smokes per day and the level of cadmium in her breast milk. A study in Japan investigated the interaction between some trace metals, including cadmium, and nutritional elements in breast milk. Researchers found a significant association between breast milk cadmium concentration and calcium secretion in breast milk. Increased cadmium in the breast milk appeared to decrease the amount of calcium secreted in the breast milk. In the study, the average concentration of cadmium in breast milk was 0.28 micrograms/liter. Cadmium toxicity targets the kidneys and bone, two crucial calcium metabolizing sites, thus decreasing the amount of calcium in blood and ultimately in breast milk. This data, in conjunction with previously noted data on blood lead levels and calcium, strongly supports previous conclusions about the role of calcium supplementation during pregnancy and lactation, and indicates that cadmium exposure may result in insufficient levels of calcium in the breast milk.

Minerals: Beneficial and Toxic

Vitamins and trace minerals are necessary components of the human diet because they are either inadequately synthesized or not synthesized in the body.

Table 2: Minerals: Beneficial and Toxic

Element	*Breast Milk Concentration ($\mu g/dL$)*	*Formula Concentration*
Average ($\mu g/dL$)		
Cobalt (Co)	0.2	0.9
Chromium (Cr)	24.3	6.9
Copper (Cu)	400	1207
Iron (Fe)	380	9227
Manganese (Mn)	6.3	46.1
Nickel (Ni)	0.8	14.3
Selenium (Se)	17	Not Quantified
Vanadium (V)	0.2	0.7
Silver (Ag)	0.4	0.7
Aluminium (Al)	67	210.5
Arsenic (As)	6.7	Not Quantified
Titanium (Ti)	6.3	13

However, only small amounts of these elements are needed to carry out the necessary biological reactions. In larger doses, the trace minerals and metals can have toxic effects. In the past, the exact concentrations of many of these trace elements in breast milk has been unclear. One study in Germany determined the concentrations of numerous essential trace elements and toxic metals in human milk and selected infant formulas. Most of the concentrations in infant formulas (except in the case of Chromium) were approximately tenfold higher than in human milk. Therefore, the level of exposure through maternal breast milk is significantly lower than the environmental exposure through infant formula. Thus the data strongly supports the role of breastfeeding in limiting the concentration of exposure to certain potentially toxic metals.

The Health Benefits of Raw Milk

There's little mention in the mainstream media these days, of traditional foods having healing properties. Sure, there's a ton of hype touting unfermented soy products, vegetable oils and supplements as modern saviours, but in reality, these items have risk-to-benefit ratios like many drugs do.

Few people are aware that clean, raw milk from grass-fed cows was actually used as a medicine in the early part of the last century. That's right. Milk straight from the udder, a sort of "stem cell" of foods, was used as medicine to treat, and frequently *cure* some serious chronic diseases. From the time of Hippocrates to until just after World War II, this "white blood" nourished and healed uncounted millions.

Clean raw milk from pastured cows is a complete and properly balanced food. You could live on it exclusively if you had to. Indeed, published accounts exist of people who have done just that. What's in it that makes it so great? Let's look at the ingredients to see what makes it such a powerful food.

Proteins

Our bodies use amino acids as building blocks for protein. Depending on who you ask, we need 20-22 of them for this task. Eight of them are considered *essential*, in that we have to get them from our food. The remaining 12-14 we can make from the first eight in the chemical factories of our bodies.

Raw cow's milk has all 8 essential amino acids, saving our bodies the work of having to convert any into usable form. About 80% of the proteins in milk are caseins-reasonably heat stable but easy to digest. The remaining 20% or so fall into the class of whey proteins, many of which have important physiological effects (bioactivity). Also easy to digest, but very heat-sensitive, these include key enzymes (specialized proteins) and enzyme inhibitors, immunoglobulins (antibodies), metal-binding proteins, vitamin binding proteins and several growth factors. Current research is now focusing on fragments of protein (peptide segments) hidden in casein molecules that exhibit anti-microbial activity.

Lactoferrin, an iron-binding protein, has numerous beneficial properties including (as you might guess) improved absorption and assimilation of iron, anti-cancer properties and anti-microbial action against several species of bacteria responsible for dental cavities. Recent studies also reveal that it has powerful antiviral properties as well.

Two other players in raw milk's antibiotic protein/enzyme arsenal are *lysozyme* and *lactoperoxidase*. Lysozyme can actually break apart cell walls of certain undesirable bacteria, while lactoperoxidase teams up with other substances to help knock out unwanted microbes too.

The immunoglobulins, an extremely complex class of milk proteins also known as *antibodies*, provide resistance to many viruses, bacteria and bacterial toxins and may help reduce the severity of asthma symptoms. Studies have shown significant loss of these important disease fighters when milk is heated to normal processing temperatures.

Carbohydrates

Lactose, or milk sugar, is the primary carbohydrate in cow's milk. Made from one molecule each of the simple sugars glucose and galactose, it's known as a disaccharide. People with lactose intolerance for one reason or another (age, genetics, etc.), no longer make the enzyme lactase and so can't digest milk sugar. This leads to some unsavory symptoms, which, needless to say, the victims find rather unpleasant at best. Raw milk, with its lactose-digesting Lactobacilli bacteria intact, may allow people who traditionally have avoided milk to give it another try.

The end-result of lactose digestion is a substance called lactic acid (responsible for the sour taste in fermented dairy products). Besides having known inhibitory effects on harmful species of bacteria, lactic acid boosts the absorption of calcium, phosphorus and iron, and has been shown to make milk proteins more digestible by knocking them out of solution as fine curd particles.

Fats

Approximately two thirds of the fat in milk is saturated. Good or bad for you? Saturated fats play a number of key roles in our bodies: from construction of cell membranes and key hormones to providing energy storage and padding for delicate organs, to serving as a vehicle for important fat-soluble vitamins.

All fats cause our stomach lining to secrete a hormone (cholecystokinin or CCK) which, aside from boosting production and secretion of digestive enzymes, let's us know we've eaten enough. With that trigger removed, non-fat dairy products and other fat-free foods can potentially help contribute to over-eating. Consider that, for thousands of years before the introduction of the hydrogenation process (pumping hydrogen gas through oils to make them solids) and the use of canola oil (from genetically modified rapeseed), corn, cottonseed, safflower and soy oils, dietary fats were somewhat more often saturated and frequently animal-based. (Prior to about 1850, animals in the U.S. were not so heavily fed corn or grain). Use of butter, lard, tallows, poultry fats, fish oils, tropical oils such as coconut and palm, and cold pressed olive oil were also higher than levels seen today.

Now consider that prior to 1900, very few people died from heart disease. The introduction of hydrogenated cottonseed oil in 1911 (as trans-fat laden Crisco) helped begin the move away from healthy animal fats, and toward the slow, downward trend in cardiovascular health from which millions continue to suffer today.

CLA, short for *conjugated linoleic acid* and abundant in milk from grass-fed cows, is a heavily studied, polyunsaturated Omega-6 fatty acid with promising health benefits. It certainly does wonders for rodents, judging by the hundreds of journal articles I've come across! There's serious money behind CLA, so it's a sure bet there's something to it.

Among CLA's many potential benefits: it raises metabolic rate, helps remove abdominal fat, boosts muscle growth, reduces resistance to insulin, strengthens the immune system and lowers food allergy reactions. As luck would have it, grass-fed raw milk has from 3-5 times the amount found in the milk from feed lot cows.

Vitamins

Volumes have been written about the two groups of vitamins, water and fat soluble, and their contribution to health. Whole raw milk has them all, and they're completely available for your body to use. Whether regulating your metabolism or helping the biochemical reactions that free energy from the food you eat, they're all present and ready to go to work for you.

Just to repeat, nothing needs to be added to raw milk, especially that from grass-fed cows, to make it whole or better. No vitamins. No minerals. No enriching. It's a complete food.

Minerals

Our bodies, each with a biochemistry as unique as our fingerprints, are incredibly complex, so discussions of minerals, or any nutrients for that matter, must deal with ranges rather than specific amounts. Raw milk contains a broad selection of completely available minerals ranging from the familiar calcium and phosphorus on down to trace elements, the function of some, as yet, still rather unclear.

A sampling of the health benefits of calcium, an important element abundant in raw milk includes: reduction in cancers, particularly of the colon: higher bone mineral density in people of every age, lower risk of osteoporosis and fractures in older adults; lowered risk of kidney stones; formation of strong teeth and reduction of dental cavities, to name a few.

An interesting feature of minerals as nutrients is the delicate balance they require with other minerals to function properly. For instance, calcium needs a proper ratio of two other macronutrients, phosphorus and magnesium, to be properly utilized by our bodies. Guess what? Nature codes for the entire array of minerals in raw milk (from cows on properly maintained pasture) to be in proper balance to one another thus optimizing their benefit to us.

Enzymes

The 60 plus (known) fully intact and functional enzymes in raw milk have an amazing array of tasks to perform, each one of them essential in facilitating one key reaction or another. Some of them are native to milk, and others come from beneficial bacteria growing in the milk. Just keeping track of them would require a post-doctoral degree! To me, the most significant health benefit derived from food enzymes is the burden they take off our body. When we eat a food that contains enzymes devoted to its own digestion, it's that much less work for our pancreas. Given the choice, I'll bet that busy organ would rather occupy itself with making metabolic enzymes and insulin, letting food digest itself.

The amylase, bacterially-produced lactase, lipases and phosphatases in raw milk, break down starch, lactose (milk sugar), fat (triglycerides) and phosphate compounds respectively, making milk more digestible and freeing up key minerals. Other enzymes, like catalase, lysozyme and lactoperoxidase help to protect milk from unwanted bacterial infection, making it safer for us to drink.

Cholesterol

Milk contains about 3mg of cholesterol per gram-a decent amount. Our bodies make most of what we need, that amount fluctuating by what we get from our food. Eat more, make less. Either way, we need it. Why not let raw milk be one source?

Cholesterol is a protective/repair substance. A waxy plant steroid (often lumped in with the fats), our body uses it as a form of waterproofing, and as a building block for a number of key hormones.

It's natural, normal and essential to find it in our brain, liver, nerves, blood, bile, indeed, every cell membrane. The best analogy I've heard regarding cholesterol's supposed causative effects on the clogging of our arteries is that blaming it is like blaming crime on the police because they're always at the scene.

Seriously consider educating yourself fully on this critical food issue. It could, quite literally, save your life.

Beneficial Bacteria

Through the process of fermentation, several strains of bacteria naturally present or added later (*Lactobacillus, Leuconostoc* and *Pediococcus,* to name a few) can transform milk into an even more digestible food.

With high levels of lactic acid, numerous enzymes and increased vitamin content, 'soured' or fermented dairy products like yogurt and kefir (made with bacteria and *yeast,* actually) provide a plethora of health benefits for the savvy people who eat them. Being acid lovers, these helpful little critters make it safely through the stomach's acid environment to reach the intestines where they really begin to work their magic

Down there in the pitch black, some of them make enzymes that help break proteins apart-a real benefit for people with weakened digestion whether it be from age, pharmaceutical side-effects or illness.

Other strains get to work on fats by making lipases that chop triglycerides into useable chunks. Still others take on the milk sugar, lactose, and, using fancy sounding enzymes like beta-galactosidase, glycolase and lactic dehydrogenase (take notes, there'll be a quiz later!), make lactic acid out of it.

As I mentioned way up yonder in the Carbohydrate section, having lactic acid working for you in your nether regions can be a good thing. Remember? It boosts absorption of calcium, iron and phosphorus, breaks up casein into smaller chunks and helps eliminate bad bugs. Raw milk is a living food with remarkable self-protective properties, but here's the kick: most foods tend to go south as they age, raw milk just keeps getting better. Not to keep harping on this, but what the heck: through helpful bacterial fermentation, you can expect an increase in enzymes, vitamins, mineral availability and overall digestibility. Not bad for old age!

A Word About Diet In General

Use common sense and stick with whole, unprocessed foods, free from genetic tweaking (there's still just too much conflicting information out there on that topic), and you'll likely be ahead of the game.

Cook your foods minimally, and you'll be even better off. Learn about sprouting and fermentation. Question everything before letting it past your lips. Explore what worked for countless generations before ours, and put it to work for yourself today. You *can* achieve great health by diet alone. I've done it, and so can you! Few people are aware that clean, raw milk from grass-fed cows was actually used as a medicine in the early part of the last century. From the time of Hippocrates to until just after World War II, this "white blood" nourished and healed millions. Clean raw milk from pastured cows is a complete and properly balanced food which you could live on exclusively if you had to. What's in it that makes it so great? Let's look at the ingredients to see what makes it such a powerful food.

Proteins: Our bodies use amino acids as building blocks for protein. Depending on who you ask, we need 20-22 of them for this task. Eight are considered essential, in that we have to get them from our food. The remaining 12-14 we can make from the first eight in the chemical factories of our bodies.

Raw cow's milk has all 20 of the standard amino acids. About 80% of the proteins in milk are caseins; reasonably heat stable but easy to digest. The remaining 20% fall into the class of whey proteins, many of which have important physiological effects (bioactivity). Also easy to digest, but very heat sensitive, these include key enzymes (specialized proteins) and enzyme inhibitors, immunoglobulins (antibodies), metal-binding proteins, vitamin binding proteins and several growth factors.

Lactoferrin, an iron-binding protein, has numerous beneficial properties including improved absorption and assimilation of iron, anti-cancer properties and anti-microbial action against several species of bacteria responsible for dental cavities. Recent studies also reveal that it has powerful antiviral properties as well.

Two other players in raw milk's antibiotic protein/enzyme arsenal are lysozyme and lactoperoxidase. Lysozyme can actually break apart cell walls of certain undesirable bacteria, while lactoperoxidase teams up with other substances to help knock out unwanted microbes.

The immunoglobulins, an extremely complex class of milk proteins known as antibodies, provide resistance to many viruses, bacteria and bacterial toxins and may help reduce the severity of asthma symptoms. Studies have shown significant loss of these important disease fighters when milk is heated to normal processing temperatures.

Carbohydrates: Lactose, or milk sugar, is the primary carbohydrate in cow's milk. Made from one molecule each of the simple sugars glucose and galactose, it's known as a disaccharide. People with lactose intolerance for one reason or another (age, genetics, etc.), no longer make the enzyme lactase and so can't digest milk sugar. This leads to some unsavory symptoms, which, needless to say, the victims find rather unpleasant at best. Raw milk, unlike pasteurized, has it's milk sugar enzyme, lactase, undamaged, and so, may allow people who traditionally have avoided milk to give it another try.

Lactose provides milk's natural complement of Lactobacilli bacteria with the raw material from which to make lactic acid (the sour taste in fermented dairy products). Besides having known inhibitory effects on harmful species of bacteria, lactic acid boosts the absorption of calcium, phosphorus and iron, and has been shown to make milk proteins more digestible.

Fats: Approximately two thirds of the fat in milk is saturated. Saturated fats play a number of key roles in our bodies: from construction of cell walls and key hormones to providing energy storage and padding for delicate organs, to providing a vehicle for carrying important fat-soluble vitamins

All fats cause our stomach lining to secrete a hormone (CCK) which, aside from boosting production and secretion of digestive enzymes, let's us know we've eaten enough. With that trigger removed, non-fat dairy products and other fat-free foods can potentially help contribute to over-eating. Consider that, for thousands of years before the introduction of the hydrogenation process (pumping hydrogen gas through oils to make them solids) and the use of canola oil (from genetically modified rapeseed), corn, cottonseed, safflower and soy oils, dietary fats were largely saturated and often animal based. Healthy cultures all over the world thrived on the use of butter, lard, tallows, poultry fats, fish oils, tropical oils such as coconut and palm, and cold pressed olive oil.

Now consider that prior to 1900, very few people died from heart disease. The introduction of hydrogenated cottonseed oil in 1911 helped begin the move away from healthy animal fats, and toward the slow, downward trend in cardiovascular health from which millions continue to suffer today.

CLA, short for conjugated linoleic acid and abundant in milk from grass-fed cows, is a polyunsaturated Omega-6 fatty acid with promising health benefits. CLA raises metabolic rate, helps remove abdominal fat, boosts muscle growth, reduces resistance to insulin, strengthens the immune system and lowers food allergy reactions. Grass-fed raw milk has 3-5 times the amount found in the milk from feed lot cows.

Vitamins: Volumes have been written about the two groups of vitamins, water and fat soluble, and their contribution to health. Whole raw milk has them all. Whether regulating your metabolism or helping the biochemical reactions that free energy from the food you eat, they're all present and ready to go to work for you.

Nothing needs to be added to raw milk to make it whole or better. No vitamins. No minerals. No enriching. It's a complete food. Heated milk must have destroyed components added back in—especially the important fat soluble vitamins A and D.

Minerals: Raw milk contains a broad selection of completely available minerals ranging from calcium and phosphorus down to trace elements.

A sampling of the health benefits of calcium, a 'macronutrient' abundant in raw milk includes: reduction in cancers, particularly of the colon; higher bone mineral density in people of every age, lower risk of osteoporosis and fractures in older adults; lowered risk of kidney stones; formation of strong teeth and reduction of dental cavities, to name a few.

An interesting feature of minerals as nutrients is the delicate balance they require with other minerals to function properly. For instance, calcium needs a proper ratio of two other macronutrients, phosphorus and magnesium, to be properly utilized by our bodies. Guess what? The entire array of minerals in raw milk is in proper balance to one another thus optimizing their benefits.

Enzymes: The most significant health benefit derived from food enzymes is the burden they take off our body. When we eat a food that contains enzymes devoted to its own digestion, it's that much less work for our pancreas. The amylase, lactase, lipase and phosphatase in raw milk, break down starch, lactose (milk sugar), fat (triglycerides) and phosphate compounds respectively, making milk more digestible and freeing up key minerals. Other enzymes, like catalase, lysozyme and lactoperoxidase help protect milk from unwanted bacterial infection.

Cholesterol: Milk contains about 3mg of cholesterol per gram—a decent amount. Our bodies make most of what we need, that amount fluctuating by what we get from our food. Cholesterol is a protective/repair substance. A waxy plant steroid (often lumped in with the fats), our body uses it as a form of water-proofing, and as a building block for key hormones. It's natural, normal and essential to find it in our brain, liver, nerves, blood, bile, indeed, every cell wall. Seriously consider educating yourself fully on this critical food issue. It could, quite literally, save your life.

Beneficial Bacteria: Through the process of fermentation, several strains of bacteria naturally present or added later (Lactobacillus, Leuconostoc and Pediococcus, to name a few) can transform milk into an even more digestible food. With high levels of lactic acid, numerous enzymes and increased vitamin content, 'soured' or fermented dairy products like yogurt and kefir (made with bacteria and yeast) provide a plethora of health benefits. Being acid lovers, these helpful little critters make it safely through the stomach's acid environment to reach the intestines where they begin to work their magic. Some of them make enzymes that help break proteins apart—a real benefit for people with weakened digestion whether it be from age, pharmaceutical side-effects or illness. Other strains get to work on fats by making lipases that chop triglycerides into useable chunks. Still others take on the milk sugar, lactose, and, using enzymes like beta-galactosidase, glycolase and lactic dehydrogenase make lactic acid out of it. Raw milk is a living food with remarkable self-protective properties. Perhaps it is time to explore all the benefits of raw milk and begin living a healthy lifestyle. Explore what worked for countless generations before ours, and put it to work for yourself today.

Milk and Milk Products

Although cow's milk is the most popular in many countries, milk can be obtained from many different sources. For example, milk from goats and sheep makes a substantial contribution to the total milk production in countries of Eastern and Southern Europe, Malawi, and Barbados, whereas the water buffalo is a common source of milk in much of Asia.

Woman Milking a Goat

Milk is a perishable commodity and spoils very easily. Its low acidity and high nutrient content make it the perfect breeding ground for bacteria, including those which cause food poisoning (pathogens).

Bacteria from the animal, utensils, hands, and insects may contaminate the milk, and their destruction is the main reason for processing. This preservation of the milk can be achieved by fermentation, heating, cooling, removal of water, and by concentration or separation of components, to produce foods such as butter or cheese. The degree to which milk consumption and processing occurs will differ from region to region. It is dependent upon a whole host of factors, including geographic and climatic conditions, availability and cost of milk, food taboos, and religious restrictions. Where processing does exist, many traditional techniques can be found for producing indigenous milk products. These are more stable than raw milk and provide a means of preservation as well as adding variety to the diet. In addition, the introduction of western-style dairy products and the subsequent setting up of small-scale dairies has provided more choice of dairy products to the consumer.

Nutritional Significance

Milk is often regarded as being nature's most complete food. It earns this reputation by providing many of the nutrients which are essential for the growth of the human body. Being an excellent source of protein and having an abundance of vitamins and minerals, particularly calcium, milk can make a positive contribution to the health of a nation. The realization of its nutritional attributes is clearly illustrated by the implementation of numerous 'school milk programmes' worldwide.

Fermented-milk products such as yoghurt and soured milk contain bacteria from the *Lactobacilli* group. These bacteria occur naturally in the digestive tract and have a cleansing and healing effect. Therefore the introduction of fermented products into the diet can help prevent certain yeasts and bacteria which may cause illness. Many people suffer from a condition known as 'lactose intolerance'. This means that they are unable to digest the milk fat (lactose). Such people can, however, tolerate milk if it is fermented to produce foods such as yoghurt. During fermentation, lactic acid producing bacteria break down lactose, and in doing so eliminate the cause of irritation.

The Quality of Milk

The type of animal, its quality, and its diet can lead to differences in the colour, flavour, and composition of milk. Infections in the animal which cause illness may be passed directly to the consumer through milk. It is therefore extremely important that quality-control tests are carried out to ensure that the bacterial activity in raw milk is of an acceptable level, and that no harmful bacteria remain in the processed products.

Table 3: Average composition (%) of milks of various mammals

Species	Water	Fat	Protein	Lactose	Ash
Human	87.43	3.75	1.63	6.98	0.21
Cow	87.2	3.7	3.5	4.9	0.7
Goat	87.00	4.25	3.52	4.27	0.86
Sheep	80.71	7.9	5.23	4.81	0.9
Indian buffalo	82.76	7.38	3.6	5.48	0.78
Camel	87.61	5.38	2.98	3.26	0.7
Horse	89.04	1.59	2.69	6.14	0.51
Llama	86.55	3.15	3.9	5.6	0.8

Standard Testing Procedures

Milk Fat

The price paid for milk is usually dependent upon the milk-fat content, and this may be determined either at the collection stage or at the dairy using a piece of equipment known as a butyrometer. Additionally the specific gravity can be measured using a hydrometer. This can also be used as an aid to detect adulteration.

Bacterial Activity

Routinely it is necessary to check the microbiological quality of raw milk using either methylene blue or resazurin dyes. These tests indicate the activity of bacteria in the milk sample and the results determine whether the milk is accepted or rejected. Both tests work on the principle of the time taken to change the colour of the dye. The length of time taken is proportional to the number of microorganisms present (the shorter the time taken, the higher the bacterial activity). It is preferable to use the resazurin test as this is less time-consuming. For these tests, basic laboratory equipment will be needed such as test-tubes, a water bath, accurate measuring equipment, and a supply of dyes. After collection the milk should ideally be stored at a temperature of 4°C or below. This is necessary to slow the growth of any contaminating bacteria.

Phosphatase Test

For pasteurized milk, it is possible to ensure that pasteurization has been adequately achieved by testing for the presence of the enzyme phosphatase. The destruction of phosphatase is regarded as a reliable test to show that the milk has been sufficiently heat-processed, because this enzyme (present in raw milk) is destroyed by pasteurization conditions. It is stressed that pasteurization is an effective safeguard against spoilage and food poisoning only if the milk is not re-contaminated after pasteurization.

Processing

Table 4: Equipment required

Processing stage	Equipment	Section reference
Store at 4°C	Refrigerated storage	15.0
	Thermometer	63.0
Test for fat content	Butyrometer	64.5
Test specific gravity	Hydrometer	64.4
Test bacterial activity	Supply of dyes ThermometerBasic laboratory equipment is required for most of the tests	64.6 63.0
Filter	Filter cloth	0.80
	Filter press	29.2
Homogenization	Homogenizer	37.0
Fill into bottles	Liquid-filling machine	28.1
	Capping machine	47.2
Pasteurization	Boiling pan	48.0
	or pasteurizer	50.0
	Heat source	36.0
	Thermometer	63.0
Sterilization	Pressure cooker	48.0
	Retort	05.1
	Heat source	36.0
	Thermometer	63.0
Cool	Bottle-cooling system	

Liquid milk

Milk can be kept for longer periods of time if it is heated to destroy the bacteria or cooled to slow their growth. Pasteurization and sterilization are the two most commonly-used heat treatments. Technically, it is possible for both to be carried out on a small scale, but they are most usually performed on a larger industrial scale due to the need for qualified, experienced staff and accurate and strictly controlled hygienic processing conditions.

Homogenization : Homogenization breaks up the oil droplets in milk and prevents the cream from separating out and forming a layer. This is of particular importance for sterilized milk which has a long shelf-life and when the formation of a cream layer is not desired. Additional changes include increased viscosity and a richer taste. Homogenizers are more usually designed for industrial-scale production, but it is possible to purchase smaller versions.

Filling : The most common packaging material for both pasteurized and sterilized milk is glass bottles sealed with either foil or metal caps, although plastic bottles, plastic bags, and cardboard cartons are all used when bottles are not available or too expensive.

Pasteurization

Pasteurization is a relatively mild heat treatment, (usually performed below 100°C) which is used to extend the shelf-life of milk for several days. It preserves the milk by the inactivation of enzymes and destruction of heat-sensitive microorganisms, but causes minimal changes to the nutritive value or sensory characteristics of a food.

Some heat-resistant bacteria survive to spoil the milk after a few days, but these bacteria do not cause food poisoning. The time and temperature combination needed to destroy 'target' microorganisms will vary according to a number of complex inter-related factors.

For milk, the heating time and temperature is either 63°C for 30 minutes or alternatively 72°C for 15 seconds. Only the former combination is possible on a small scale and for this the simplest equipment required is an open boiling pan. Better control is achieved using a steam jacketed pan, and this can be fitted with a stirrer to improve the efficiency of heating. Both of these are batch processes which are suited to small-scale operation. A higher production rate may be possible using a tubular-coil pasteurizer. This equipment has been tested and has been successful for some fruit products but it is presently still at a developmental stage.

Sterilization

Sterilization is a more severe heat treatment designed to destroy all contaminating bacteria. The milk is sterilized at a temperature of 121°C maintained for 15-20 minutes. This can be achieved using a retort or pressure cooker. Unlike pasteurization, this process causes substantial changes to the nutritional and sensory quality of the milk. In some countries, flavoured milk has become a very popular product.

However, sterilization is not recommended for small-scale production for the following reasons:

- The cost of a retort and ancillary equipment is high for the small-scale processor.
- It is essential that the correct heating conditions are carefully established and maintained for every batch of milk that is processed. If the milk is overheated, the quality is reduced, and it may have a rather burnt taste and aroma.
- If the milk is not heated sufficiently, there is a risk that microorganisms will survive and grow inside the bottle. In low-acid foods such as milk, many types of bacteria including *Clostridium botulinum* can grow and cause severe food poisoning.
- Due to the potential dangers from food poisoning, the skills of a qualified food technologist/microbiologist are required in order to routinely examine samples of sterilized milk that have been subjected to accelerated storage conditions. This requires a supply of microbiological media and equipment.

In summary, the process of sterilization requires a considerable capital investment, the need for trained and experienced staff, regular maintenance of sophisticated equipment, and a comparatively high operating expenditure.

Cooling

Pasteurization does not destroy all of the microorganisms, therefore the milk has to be cooled rapidly to prevent the growth of surviving bacteria. Cooling can be achieved on a small scale by using a bottle-cooling system.

Storage

Pasteurized milk has a shelf-life of 2-3 days if kept at 4°C. Maintaining this low temperature causes a substantial increase to the cost of transportation and distribution and is therefore a major disadvantage to the development of a small-scale pasteurized milk business. If packaged in sealed bottles and stored at room temperature, sterilized milk should have a shelf-life in excess of six months.

Separation of Milk Components

Cream

When milk is left to stand for some time, fat globules rise to the surface forming a layer of fat (or cream). This can be separated leaving behind skimmed milk as a by-product. There are different types of cream each with different fat concentrations: single (or light) cream contains 18 per cent milk fat whereas double (or heavy) cream normally contains 30 per cent milk fat. Cream is a luxury item and may be used as an accompaniment to coffee, as a filling in cakes, and an ingredient in ice cream.

Separation

Separation can very simply be achieved by removing the cream with a spoon, however this is a slow process during which the cream may spoil. For this reason it is more usual to use a manual or powered centrifuge.

Table 5: Production stages for cream

Ingredients	Process	Equipment	Section reference
Raw milk tested	Store at 4°C	Milk churns	62.0
	Refrigerated storage	15.0	
	Thermometer	63.0	
	Separation of milk fat	Ladle Dairy centrifuge	07.1
	Pasteurization	Large boiling pan or steam jacketed pan	48.0
	Pasteurizer	50.0	
	Heat source	36.0	
	Thermometer	63.0	
	Fill bottles/pots	Funnel or liquid-filling machine	28.1
	Capping machine	47.2	
	Pot sealer	47.1	
	Cool bottles	Bottle-cooler	
	Store bottles at 4°C	Refrigerated storage	15.0

Pasteurization

Cream may be pasteurized in a similar way to milk, using a similar time and temperature combination and the same equipment. Cream can also be sterilized but there is a considerable loss of quality.

Packaging and Storage

Cream can be packaged in glass jars or plastic pots sealed with foil lids. Pasteurized cream must be stored at a temperature of 4°C to have a shelf-life of several days. Refrigerated storage is necessary because cream is prone to rapid spoilage.

Churning

Churning disrupts the emulsion of fat and water and as a result the milk-fat separates out into granules. This process takes place in a butter churn.

Churning Cream

Churning is continued until fat granules are present and at this stage the mixture is drained to remove liquid that has separated from the granules. This liquid is known as buttermilk and can be used as either a beverage or as an ingredient in animal feed.

Washing

Clean water equivalent in weight to the buttermilk is added to the churn in order to wash the butter granules. The wash water is drained away. Churning is continued for a short time to compact the butter, and once this has been achieved it is removed from the churn.

Forming and Packaging

Butter is kneaded to achieve a smooth and pliable texture. This can be done using simple hand-tools such as butter pats. Alternatively for higher production rates a specially-designed kneader can be used. Once the butter has a uniform and smooth texture it is formed into blocks with butter pats and packed in either greaseproof paper or foil wrappers.

Working Butter with Butter Pats

Storage

Due to its high fat composition, butter must be stored at temperatures below 10°C otherwise the fat becomes rancid and imparts undesirable 'off' flavours. The water droplets in butter (20 per cent) can also allow bacteria to grow if it is not kept under cool conditions.

Ghee

Ghee is made from butter which has been heated and clarified. At ambient temperatures it is a semi-solid mass with a granular texture, but on melting (40°C+) it turns into a clear, thin liquid. It has a high demand in some countries for domestic use, as an ingredient for local food production (for example bakeries and confectionery manufacturers), and as an export commodity.

Alternatively, cream is boiled gently to evaporate the water. During boiling the product is stirred continuously until the milk proteins start to coagulate, forming particles, and the colour of the cream darkens. Heating is stopped and the product is left to set. The particles settle at the bottom of the vessel and the milk-fat is separated. The principles of preservation are:

- heating to destroy enzymes and contaminating microorganisms
- to reduce the water-content by evaporation, and in doing so prevent the growth of microorganisms.

Packaging and Storage

Metal containers are normally used. They should be thoroughly cleaned, especially if they are re-usable, and they should be made airtight. Alternatives to metal cans include coloured glass jars with metal lids, or ceramic pots sealed with cork/plastic stoppers.

Ghee is usually stored at room temperatures as cold storage affects the granular texture. Thus ghee is useful for those consumers with no access to refrigeration.

Cultured/Fermented Dairy Products

The technology of cultured milk products such as yoghurt, curd, and cheese is based upon the microbial conversion of the milk-sugar lactose to lactic acid (lactic acid accounts for the characteristic 'sourness' of such products). In order for the conversion to take place, lactic acid producing bacteria must be present. This may occur by allowing the milk to sour naturally, but it is better to introduce the appropriate bacteria as a starter culture. Starter cultures may be in the form of a small quantity of previously-cultured product or may be purchased as a commercially-prepared culture.

Yoghurt/Curd

Yoghurt is a fermented milk product that evolved by allowing naturally-contaminated milk to sour at a warm temperature. Yoghurt can be either unsweetened or sweetened, set, or stirred. Curd is the name given to a yoghurt-type product made from buffalo milk.

The principles of preservation for yoghurt are:
- Pasteurization of the raw milk to destroy contaminating microorganisms and enzymes.
- An increase in acidity due to the production of lactic acid from lactose. This inhibits the growth of food-poisoning bacteria.
- Storage at a low temperature to inhibit the growth of microorganisms.

Heating

In the manufacture of yoghurt, milk is normally heated to 70°C for 15-20 minutes, using an open boiling pan, or alternatively a steam jacketed pan.

Addition of Starter Culture

The milk is cooled to between 30 40°C and inoculated with a mixed culture of *Lactobacillus bulgaricus* and *Streptococcus thermophilus* (usually in a ratio of 1:1). If a commercial starter-culture is used, the directions for use will be given. However, if a culture from a previous batch is used, then it is usual to add 2-3 tablespoons per litre of prepared milk. Yoghurt of the stirred variety can be fermented in the mixing container. To make set yoghurt the inoculated milk should be poured into the individual pots before fermentation.

Selling Curd from a Roadside Stall

Incubation

The microorganisms that produce yoghurt are most active within a temperature range of 32-47°C. Ambient temperatures are therefore not adequate and a heated incubator is needed. Small commercially-available yoghurt-makers consist of an electrically-heated base and a set of plastic or glass containers. Most yoghurt-makers make four or five individual half litre cups at a time.

There are other simple and inexpensive ways of incubating yoghurt such as an insulated box, keeping the jars/pots surrounded by warm water, or by using thermos flasks (the latter is only suitable for stirred yoghurt). Incubation takes approximately five hours. When fermentation is complete, stirred yoghurt is cooled and flavoured or sweetened prior to packaging. In set yoghurt all ingredients are added before fermentation.

Packaging and Storage

Yoghurt or curd is commonly packaged in plastic pots fitted with a plastic lid, or heat-sealed with foil, although traditionally, curd is packaged in clay pots. Such pots are made from local materials and can be re-used or later used for cooking. The shelf-life of yoghurt is usually 3-8 days when stored at temperatures below 10°C.

Cheese

Cheese is made from milk by the combined action of lactic acid bacteria and the enzyme rennin (known as rennet). Just as cream is a concentrated form of milk fat, cheese is a concentrated form of milk-protein. The differences in cheeses that are produced in different regions result from variations in the composition and type of milk, variations in the process, and the bacteria used. The different cheese varieties can be classified as either hard or soft.

Bibliography

Alec, William: *The Dairy Chemical Industry*, London: Longman Group Limited, 1971.

Anifantakis, EM.: *Greek Cheeses*, Nat. Dairy Committee of Greece Publ., Athens, Greece, 1991.

Arora, Dinesh: *Biotech's Dictionary of Dairy Science*, Biotech Books, Delhi, 296.

Ashby, E.: *Technology and the Academics*, London: Macmillan, 1959.

Barbeau, G.: *Tropical Fruits in Nicaragua*, Managua, Nicaragua Ministerio de Desarrollo Agropecuario, Agraria, 1990.

Basavaraj, S. Benni: *Dairy Co-operative Management and Practice*, Rawat, Delhi, 2005.

Bhattacharya, Lata: *Biochemistry of Nutrition*, Discovery, Delhi, 2010.

Bohra, Babita: *Dairy Farming in Mountain Areas*, Daya, Delhi, 2006.

Brock, H.: *History of Dairy Chemistry*, New York: Norton, 1992.

Bucciarelli, L.: *Designing Engineers in Dairy*, Cambridge: MIT Press, 1995.

Bushnell, R.B.: *Dry Cow Feeding and Management*, A Western Regional Extension Publication, 1979.

Chand, Ram: *Decision-Making of Dairy Beneficiaries: Role of Aspiration, Motivation and Knowledge*, Om Pub, Delhi, 2010.

Chhazllani, V K: *Dairy Chemistry and Animal Nutrition*, Manglam Pub, Delhi, 2008.

Cohen, Lizabeth, *Making a New Deal, Industrial Workers in Chicago, 1919-1939* Cambridge University Press, 1991.

Damasio, AR.: *Descarte's Error: Emotion, Reason and the Human Brain*, New York, 1994.

David, M.: *Ideas in Chemistry: A History of the Science*, New Brunswick, N.J.: Rutgers University Press, 1992.

De, Sukumar: *Outlines of Dairy Technology*, Oxford University Press, Delhi, 2001.

Deepti, R. Haritash and Tej Verma: *Promoting Community Nutrition Through Participatory Monitoring and Evaluation*, Advance, 2001.

Desmond, Glenn, and John A, Marcell, *Handbook of Small Business Valuation Formulas and Rules of Thumb,* Valuation Press, 1993.

Devraj, B: *Impact of Dairy On Small and Marginal Farmers,* Prateeksha Publications, Delhi, 2010.

Droop, H. Richmond: *Laboratory Manual of Dairy Analysis,* Biotech, 2004.

Fox, Patrick F.: *Advanced Dairy Chemistry: Proteins,* New York: Elsevier Applied Science, 1992.

Friberg, Stig. E..: *Food Emulsions,* New York: M. Dekker, 1997.

Gautam, U.S. and Ram Chand: *Decision-Making of Dairy Beneficiaries: Role of Aspiration, Motivation and Knowledge,* Om Pub, Delhi, 2010.

Gowdy, J., *Coevolution Economics: The Economy, Society and the Environment,* Kluwer, Boston, 1994.

Guarti, Luigi, *The Valuation of Firms,* Blackwell Publishing, 1994.

Gupta, Sudhir: *Hand Book of Dairy Formulations, Processes and Milk Processing Industries,* Engineers India Res Inst, Delhi, 2003.

Hemantaranjan, A.: *Advancements in Iron Nutrition Research,* Scientific, Delhi, 1995.

Hornig, Susanna: *A Grain of Truth: The Media, the Public, and Biotechnology,* Lantham, MD: Rowman and Littlefield, 2001.

Hui, Yiu H.: *Dairy Science and Technology Handbook,* New York: Wiley, 1993.

Jensen, Robert G.: *Handbook of Milk Composition,* San Diego: Academic Press, 1995.

Jensen, Robert G.: *Handbook of Milk Composition,* San Diego: Academic Press, 1995.

Jha, S. N.: *Dairy and Food Processing Plant Maintenance: Theory and Practice,* International Book Distributing, Delhi, 2006.

John A, Marcell, *Handbook of Small Business Valuation Formulas and Rules of Thumb,* Valuation Press, 1993.

Julia, F.: *Fruits of Warm Climates,* Miami, Julia F. Morton Publisher, 1987.

Kango, Mangala: *Normal Nutrition: Fundamental and Management,* RBSA, Delhi, 2003.

Kapoor, Ajay: *Dairy Science and Technology,* Vishvabharti Pub, Delhi, 2005.

Khemka, Devyani: *Animal Physiology,* Dominant, Delhi, 2003.

King, C.J.: *Separation Processes,* McGraw Hill, 1980.

Koli, P. A.: *Dairy Development in India: Challenges Before Co-Operatives,* Shruti Pub, Delhi, 2007.

Kosikowski, Frank, and Vikram V. Mistry: *Cheese and FermentedMilk Foods,* Great Falls, Va.: Kosikowski, 1997.

Lata Bhattacharya: *Biochemistry of Nutrition,* Discovery, Delhi, 2010.

Law, Barry A.: *Microbiology and Biochemistry of Cheese and Fermented Milk*, London: Blackie Academic & Professional, 1997.

March, J. G.: *A Behavioural Theory of the Firm*, Englewood-Cliffs, Prentice Hall, 1963.

Marth, Elmer H., and James L. Steele: *Applied Dairy Microbiology*, New York: M. Dekker, 1998.

Mayrose, V.B.: *Handbook of livestock Management Techniques*, Macmillan Publishing Co., New York, 1981.

Meyer, M. W.: *Theory of Organizational Structure*, Indianapolis, 1977.

Michael, J. Lewis,: *SeparationProcesses in the Food and Biotechnology Industries*, Cambridge, U.K.: Woodhead, 1996.

Miller, Gregory D., Judith K. Jarvis, and Lois D. McBean.: *Handbook of Dairy Foods and Nutrition*, Boca Raton, Fla.: CRC Press, 1999.

Mowery, D. C.: *Technology and Wealth of Nations*, Stanford: Stanford University Press, 1992.

Mudgal, V. D.; K. K. Singhal and D. D. Sharma: *Advances in Dairy Animal Production*, International Book Distribut, 2003.

Myers, R.: *Basics of Dairy Chemistry*, Atlantic, Delhi, 2007.

Nedderman, R. M.: *A Handbook of Unit Operations*, Academic, London, 1971.

Nisha, Maimun: *Health, Food and Nutrition*, Kalpaz, Delhi, 2006.

Pal, Kanwar Singh Sangwan: *Technology of Dairy Plant Operations*, Agrobios, Delhi, 2008.

Pandey, D. N. and Amita Bajpai: *Recent Trends in Animal Nutrition and Feed Technology for Livestock, Pets and Laboratory Animals*, International, 2003.

Parihar, Pradeep and Leena Parihar: *Dairy Microbiology*, Agrobios, Delhi, 2008.

Patton, Stuart: *Principles of Dairy Chemistry*, Huntington, N.Y.: Krieger, 1976.

Pirtle, Thomas Ross: *History of the Dairy Industry*, Chicago: Mojonnier Bros. Company, 1926.

Porter, A.R., J.A. Sims, and C.F. Foreman: *Dairy Cattle in American Agriculture*, Iowa, Iowa State University Press, 1965.

Prince, John: *Dairy Farming: Being the Theory, Practice, and Methods of Dairying*, New York, 1888.

Qystein, V. Sjaastad: *Physiology of Domestic Animals*, International Book Distributing Co., Delhi, 2005.

Rao, M K: *Food and Dairy Microbiology*, Manglam Pub, Delhi, 2007.

Rao, P. Venkateshwara: *Dairy Farm Business Management*, Biotech Books, Delhi, 2008.

Samvel, A. P. V.: *Agri-Business Management*, Satish Serial Pub, Delhi, 2008.

Sarkar, A: *Advanced Organic Chemistry: Reactions and Mechanisms*, Swastik Publications, Delhi, 2011.

Sharma, Ramakant: *Chemical and Microbiological Analysis of Milk and Milk Products*, International Book Distributing, 2006.

Sheldon, John Prince: *Dairy Farming: Being the Theory, Practice, and Methods of Dairying*, New York, 1888.

Shukla, Arvind N.: *Textbook of Dairy Chemistry*, Discovery Pub, Delhi 2010.

Shukla, N.: *Textbook of Dairy Chemistry*, Discovery Pub, Delhi 2010.

Singh, G.: *Chemistry of Amino-Acids and Proteins*, Discovery, Delhi, 2007.

Singh, Harmeet: *Dairy Farming*, APH, Delhi, 2005.

Singh, Vir and Babita Bohra: *Dairy Farming in Mountain Areas*, Daya, Delhi, 2006.

Singhal, K. K. and D. D. Sharma: *Advances in Dairy Animal Production*, International Book Distribut, 2003.

Spreer, Edgar: *Milk and Dairy Product Technology*, Translated by Axel Mixa. New York: M. Dekker, 1998.

Sukumar De: *Outlines of Dairy Technology*, Oxford University Press, Delhi, 2001.

Suresh, M: *Dairy Development and Income Distribution in India*, Abhijeet Pub, Delhi, 2007.

Taylor, F. W.: *The Principles of Scientific Management*, New York, 1917.

Thompson, Paul B.: *Food Biotechnology in Ethical Perspective*, Aspen, CO: Aspen Publishers, 1997.

Thomson, Sutherland: *Grading Dairy Produce*, Medi World Press, Delhi, 1995.

Tomar, S.K. and Gunjan Goel: *Applied Dairy and Food Microbiology*, Agrotech, 2005.

Tyagi, Prasum: *A Textbook of Animal Physiology*, Dominant, Delhi, 2010.

Upadhyay, K.G. and Vyas, S.H.: *Composition of Camel's Milk*, Gujarat Agric. University, 1982.

van, M.A. J. S. Boekel: *Dairy Technology: Principles of Milk Properties and Processes*, New York: M. Dekker, 1999.

Walstra, Pieter, and Robert Jenness: *Dairy Chemistry and Physics*, New York: Wiley, 1984.

Weimar, Mark R. and Don P. Blayney: *Landmarks in the U.S. Dairy Industry*, USDA, ERS, Agriculture Information Bulletin Number 694, 1994.

Yegge, Wilbur M., *A Basic Guide for Valuing a Company*, New York, Wiley, 1996.

Index

A

Alcohol Test, 42, 55, 216, 217.
Alcohol-Alizarin Test, 42.

B

Bacteria, 2, 3, 6, 12, 18, 19, 20, 21, 22, 24, 25, 29, 38, 42, 47, 48, 50, 52, 59, 60, 71, 73, 74, 75, 77, 78, 84, 87, 88, 92, 114, 121, 124, 128, 133, 134, 135, 136, 137, 159, 160, 169, 186, 187, 195, 197, 198, 205, 207, 208, 232, 234, 235, 247, 249, 250, 252, 254, 255, 256, 257, 258, 259, 261, 262, 263, 264.
Bacteriological Tests, 38.
Bacteriophage, 19, 29.
Beverage Milks, 128.
Biochemistry, 249.
Biosynthesis, 5.
Breeds, 7, 174, 177, 182.
Bulk Starter, 29.
Butter, 6, 9, 10, 11, 12, 13, 18, 36, 75, 79, 81, 83, 84, 94, 100, 102, 106, 107, 108, 109, 110, 111, 112, 113, 114, 123, 124, 125, 127, 128, 131, 139, 149, 158, 160, 167, 168, 175, 177, 178, 179, 180, 209, 211, 223, 224, 248, 253, 255, 261, 262.
Buttermilk, 12, 13, 107, 108, 109, 111, 125, 131, 133, 140, 151, 156, 215, 218, 219, 261.

C

Cadmium, 237, 238, 239, 240, 241, 244, 245.
Campylobacter Jejuni, 26, 34.
Carbohydrates, 25, 148, 156, 210.
Casein, 13, 18, 51, 55, 85, 86, 87, 89, 90, 91, 92, 93, 114, 115, 119, 123, 125, 126, 132, 133, 135, 176, 179, 180, 183, 184, 185, 186, 187, 188, 189, 190, 192, 193, 195, 196, 199, 200, 211, 213, 246, 251.
Cattle, 6, 7, 19, 174, 224, 227, 228, 229.
Cheese, 6, 9, 10, 11, 12, 13, 18, 19, 25, 28, 29, 36, 47, 50, 57, 75, 78, 82, 87, 88, 90, 91, 92, 93, 94, 102, 115, 116, 117, 118, 119, 120, 121, 123, 124, 125, 126, 132, 133, 134, 135, 136, 137, 174, 187, 194, 195, 196, 197, 198, 199, 200, 201, 207, 208, 211, 221, 228, 229, 234, 235, 255, 262, 264.
Cheese Milk, 134.
Chocolate Milk, 128.
Cholesterol, 205, 230, 250, 254.
Churning Cream, 108, 261.
Cream Production, 209.
Curd, 13, 36, 37, 45, 90, 93, 94, 111, 112, 114, 116, 117,

118, 119, 126, 134, 135, 136, 186, 194, 195, 196, 197, 217, 247, 262, 264.

D

Dairy, 5, 6, 7, 8, 9, 10, 11, 13, 17, 18, 19, 22, 23, 24, 25, 26, 28, 29, 36, 38, 39, 44, 52, 54, 56, 57, 61, 64, 74, 75, 78, 94, 96, 97, 112, 128, 129, 130, 134, 141, 142, 150, 151, 171, 172, 207, 208, 210, 211, 217, 218, 222, 233, 234, 237, 238, 239, 241, 242, 252, 255, 257, 258, 260, 261, 262.
Dairy Foods, 165.
Dairy Industry, 7, 8, 9, 18, 39, 96, 111, 127, 165, 205.
Dairy Processing, 9, 94, 96.
Dilution, 33, 59, 69.
Diseases, 19.
Diseases, 2, 3, 19, 26, 202, 246.

E

Enzyme Coagulation, 135.
Enzymes, 5, 25, 26, 48, 55, 74, 83, 84, 87, 90, 91, 92, 115, 117, 135, 136, 174, 175, 197, 198, 201, 204, 230, 246, 248, 249, 250, 251, 252, 254, 258, 262, 263.
Equilibria, 89.

F

Fatty Acids, 51, 55, 83, 84, 176, 177, 199, 236, 237, 238, 239, 240, 241.
Feeding, 7, 16, 18, 47, 226, 227, 228, 229, 231.
Fermented Milks, 94, 121, 122, 123.
Fertilizers, 228.

Fluid Milk Processing, 127.
Food Production, 61, 62, 63, 262.
Food Quality, 30, 31.
Freezing Point Depression, 144, 153, 163.
Frozen Yogurt, 139, 147.
Fruit Ice Cream, 168.

G

Gerber Butterfat Test, 44.

H

HACCP Application, 65.
Heat Stress, 226.
Heat Treatment, 11, 42, 50, 51, 52, 54, 56, 74, 93, 130, 131, 132, 133, 134, 152, 160, 207, 258, 259.
Hydrolytic Rancidity, 84, 134.

I

Ice Cream, 12, 36, 75, 78, 88, 127, 130, 131, 141, 142, 143, 144, 145, 146, 148, 149, 150, 151, 152, 153, 154, 155, 156, 159, 160, 161, 162, 163, 164, 165, 166, 167, 168, 169, 170, 171, 172, 173, 205, 209, 210, 235, 260.

K

Kulfi, 78.

L

Lactase, 88, 175, 205, 247, 249, 252, 254.
Lactic Acid Bacteria, 24, 29, 47, 134, 208, 264.
Lactobacillus, 25, 26, 28, 137, 140, 187, 205, 206, 207, 208, 250, 254, 263.
Lactoglobulin, 56, 85, 86, 91, 93, 131, 179.

Index

Lactometer Test, 46.
Lactose, 13, 24, 25, 28, 54, 55, 58, 59, 61, 87, 88, 91, 114, 120, 125, 130, 131, 132, 133, 134, 136, 137, 141, 144, 148, 150, 151, 152, 154, 174, 176, 183, 186, 193, 194, 203, 205, 207, 208, 209, 224, 229, 247, 249, 250, 252, 254, 255, 256, 262, 263.
Lactose Intolerance, 205, 247, 252.
Leuconostoc, 25, 28, 250, 254.
Lipolytic, 26, 28.
Livestock, 18.

M

Mercury, 237, 238, 239, 240, 241, 243, 244.
Metabolism, 4, 20, 23, 74, 208, 232, 248, 253.
Microbiology, 1, 2, 3, 4, 5, 31, 32, 135.
Microorganisms, 1, 2, 3, 4, 5, 6, 19, 20, 21, 22, 23, 24, 25, 26, 28, 29, 51, 57, 72, 73, 74, 77, 121, 122, 124, 130, 131, 138, 201, 202, 207, 256, 258, 259, 262, 263.
Milk Consumption, 128, 255.
Milk Handling, 203.
Milk Lipids, 230, 231.
Milk Production, 11, 15, 26, 36, 39, 65, 109, 131, 132, 205, 230, 244, 264.
Minerals, 22, 92, 133, 150, 203, 209, 224, 229, 245, 248, 249, 253, 254, 255.
Molds, 2, 164, 194, 195, 196, 197, 198, 199, 208.

N

Nutrition, 85, 229.

O

Oxidation, 51, 83.

P

Pasteurization, 3, 26, 50, 51, 57, 60, 61, 72, 73, 74, 93, 127, 131, 132, 133, 134, 138, 139, 155, 159, 160, 195, 198, 201, 202, 203, 207, 257, 259.
Pathogenic Microorganisms, 3, 25.
Pathogens, 24, 25, 26, 36, 37, 50, 57, 71, 112, 122, 124, 160, 201, 207, 208, 255.
Probiotics, 6.
Protein, 25, 51, 52, 53, 55, 56, 58, 59, 85, 86, 87, 88, 89, 90, 91, 92, 93, 119, 126, 132, 133, 135, 136, 137, 144, 150, 151, 152, 155, 159, 161, 167, 170, 174, 181, 182, 186, 187, 191, 192, 193, 194, 196, 197, 200, 203, 204, 205, 209, 224, 225, 226, 227, 228, 229, 233, 234, 246, 247, 251, 252, 255, 264.
Public Health, 40, 202.

R

Record Keeping, 27, 64.
Reproduction, 21, 29, 85.
Resazurin Test, 43, 44.
Ripening Period, 37, 125.

S

Sedimentation Test, 52.
Sorbet, 147.
Spoilage, 4, 19, 23, 24, 25, 26, 51, 52, 74, 76, 113, 121, 122, 131, 134, 159, 201, 207, 208, 235, 257, 261.
Spoilage Microorganisms, 131, 201.
Starter Bacteria, 208.
Starter Culture, 29, 137, 263.
Starter culture, 28, 29, 47, 48, 135, 137, 138, 140, 262.

Sterilization, 50, 51, 129, 130, 258, 259.
Storage Temperature, 56, 75, 121.
Streptococcus, 25, 26, 28, 137, 187, 206, 207, 208, 263.

T

Temperature, 21, 23, 36, 46, 47, 48, 50, 51, 52, 54, 55, 56, 58, 69, 72, 75, 76, 78, 83, 84, 92, 93, 94, 95, 97, 99, 107, 108, 109, 115, 116, 121, 122, 123, 124, 125, 134, 136, 138, 139, 140, 141, 147, 149, 151, 154, 160, 161, 162, 163, 171, 172, 173, 177, 178, 179, 180, 181, 182, 183, 185, 187, 188, 189, 190, 191, 193, 194, 196, 197, 198, 199, 200, 201, 202, 204, 212, 214, 217, 218, 221, 222, 235, 257, 258, 259, 260, 261, 262, 263.
Treatment, 4, 6, 11, 25, 42, 50, 51, 52, 54, 55, 56, 74, 88, 93, 95, 123, 129, 130, 131, 132, 133, 134, 135, 139, 152, 160, 182, 194, 207, 258, 259.
Tuberculosis, 3, 26, 50, 71.

V

Viable Enumeration, 24.
Vitamins, 22, 83, 89, 133, 174, 175, 229, 230, 247, 248, 251, 252, 253, 255.

W

Whey, 9, 12, 13, 55, 78, 79, 85, 86, 88, 90, 91, 92, 93, 115, 116, 117, 118, 119, 120, 121, 122, 123, 125, 126, 131, 132, 133, 135, 136, 137, 138, 148, 150, 151, 157, 159, 175, 187, 196, 197, 199, 206, 210, 218, 219, 246, 251.
Whey Protein Concentrate, 137.

Y

Yeasts, 19, 21, 22, 35, 136, 235, 255.
Yoghurt, 6, 36, 50, 79, 82, 137, 255, 256, 262, 263, 264.

❏❏❏